1990

Titles in This Series

Volume

Titles in This Series

Titles in This Series

Volume

63 **Integral geometry,** Robert L. Bryant, Victor Guillemin, Sigurdur Helgason, and R. O. Wells, Jr., Editors

64 **The legacy of Sonya Kovalevskaya,** Linda Keen, Editor

65 **Logic and combinatorics,** Stephen G. Simpson, Editor

66 **Free group rings,** Narian Gupta

67 **Current trends in arithmetical algebraic geometry,** Kenneth A. Ribet, Editor

68 **Differential geometry: The interface between pure and applied mathematics,** Mladen Luksic, Clyde Martin, and William Shadwick, Editors

69 **Methods and applications of mathematical logic,** Walter A. Carnielli and Luiz Paulo de Alcantara, Editors

70 **Index theory of elliptic operators, foliations, and operator algebras,** Jerome Kaminker, Kenneth C. Millett, and Claude Schochet, Editors

71 **Mathematics and general relativity,** James A. Isenberg, Editor

72 **Fixed point theory and its applications,** R. F. Brown, Editor

73 **Geometry of random motion,** Rick Durrett and Mark A. Pinsky, Editors

74 **Geometry of group representations,** William M. Goldman and Andy R. Magid, Editors

75 **The finite calculus associated with Bessel functions,** Frank M. Cholewinski

76 **The structure of finite algebras,** David C. Hobby and Ralph Mckenzie

77 **Number theory and its applications in China,** Wang Yuan, Yang Chung-chun, and Pan Cheng-biao, Editors

CONTEMPORARY MATHEMATICS

Volume 77

Number Theory and Its Applications in China

Wang Yuan, Yang Chung-chun, and Pan Chengbiao, Editors

AMERICAN MATHEMATICAL SOCIETY

Providence · Rhode Island

1980 *Mathematics Subject Classification* (1985 *Revision*). Primary 11.

Library of Congress Cataloging-in-Publication Data

Number theory and its applications in China/Wang Yuan, Yang Chung-chun, and Pan Cheng-biao, editors.

p. cm. – (Contemporary mathematics, ISSN 0271-4132; v. 77)

Bibliography: p.

ISBN 0-8218-5084-9 (alk. paper)

1. Numbers, Theory of–Research–China. I. Wang, Yüan, fl. 1963-. II. Yang, Chung-Chun, 1942– III. Pan, Cheng-biao. IV. Series: Contemporary mathematics (American Mathematical Society); v. 77.

QA241.N8673 1988 88-19373
512′.7–dc 19 CIP

Printed in the United States of America.
This volume was printed directly from author-prepared copy.
The paper used in this book is acid-free and falls within the guidelines established to ensure permanence and durability. ∞

Dedicated to the memory of

Hua Lookeng 华罗庚

—— a superb researcher and a great teacher

CONTENTS

PREFACE

Number theory is one of the earliest explored modern mathematical fields in China. Undoubtedly, among all the mathematical disciplines pursued by Chinese, it is also the most accomplished one.

It was first introduced into China in the early 1920's by Yang Wu-zhi. The most important pioneers in the development of this field in China are Hua Loo-keng, Ko Chao, and Min Si-he. Particularly noted is Hua, who, from 1936-1938, visited Cambridge University of England and studied under G.H. Hardy. His works in the estimation of complete exponential sums, Warings problems, Tarry's problems, and Vinogradov's method constitute a significant contribution to the development of number theory. These accomplishments also caused him to be regarded by his contemporaries as one of the world's leading number theorists of his generation.

After the founding of the People's Republic of China, Hua was appointed as the director of the Institute of Mathematics, Academia Sinica. There, he organized two seminars: one on Goldbach's problem and the other on elementary number theory. Min Si-he organized a seminar on analytic number theory at Peking University. Ko Chao led a team doing reserach on Diophantine equations at Si Chuan University. As a result of these events, many competent young mathematicians came to the fore and the most noted accomplishments by Chinese mathematicians were focused toward the solution of the Goldbach conjecture and on sieve method. For instance, based on the work of Wang Yuan and Pan Cheng-dong, Chen Jing-run was able to prove that every sufficiently large even number can always be expressed as the sum of a prime number and an integer having at most two prime factors.

Unfortunately, the cultural revolution interrupted the research of number theory in China for more than ten years. After that the study of number theory has been resumed. Now besides Beijing and Si Chuan, the Shandong University and Chinese University of Science and Technology also have advanced programs in number theory. More specifically, in addition to the old topics, a wide variety of subjects in diophantine approximations, transcendental number theory, uniform distribution, modular forms, and algebric number theory have been taken up by middle age scholars. Chinese number theorists also have been interested in the applications of number theory.

This book contains nine survey articles and three articles of authors on recent research work. It mainly covers research topics relating: the circle method and the methods of estimating the exponential sums, Waring's problem and Goldbach's problem, weighted sieve method and mean value theorems, Riemann zeta function, arithmetic function, Dirichlet L-functions, number theoretic methods in numerical analysis, diophantine equations and diophantine inequalities in algebraic number fields, diophantine approximations and transcendental number theory, modular forms, and applications of number theory to digital signal processing and public-key systems.

Notice that in this collection we emphasize the introducing and gathering of research accomplishments of Chinese number theorists during 1949-1979, a period of time during which correspondence between Chinese and foreigners was discouraged. We also remark that here and in the text, the Chinese authors' names are spelled as they should be called in China by putting the surname before the first name.

The contributors to this volume are leading experts in China. It is hoped that the collection will not only survey the significant contributions of Chinese mathematicians, but also reflect broadly the latest developments and current state of the research of number theory in China. Thus a more concrete scientific exchange between East and West in this area will result.

We are grateful to W. Adam, D. Hejhal, C. Osgood, J. Hsia, and K. Shiota for their reviewing work, corrections ana comments for the articles. We would like to especially mention Hejhal who voluntarily reviewed the entire manuscript with great patience and enthusiasm and made many final corrections and useful suggestions for improving the text. We would also like to take this opportunity to thank M. Lane, publication director of the American Mathematical Society (AMS), I. Kra, chairman of the editorial board for Contemporary Mathematics Series of AMS for their endorsements and help provided for this project, and V. Sauber for her excellent typing. Finally, but not the least, we thank all the contributors for their participation and cooperation in the making of this book.

Wang Yuan 王元
Yang Chung-chun 杨重骏
Pan Chengbiao 潘承彪

CONTRIBUTORS

Chen, Jingrun 陈景润
Institute of Mathematics, Academia Sinica, Beijing, China

Feng, Xuning 冯绪宁
Institute of Applied Mathematics, Academia Sinica, Beijing, China

Lai K. F. 黎景辉
The Department of Mathematics, Chinese University, Hong Kong

Li, Delang 李德琅
The Department of Mathematics, Sichuan University, Chengdu, China

Liu, Mingchit 廖明哲
The Department of Mathematics, Hong Kong University, Hong Kong

Lu, Hongwen 陆洪文
The Department of Mathematics, University of Science and Technology of China, Hefei, China

Pan, Chengbiao 潘承彪
The Department of Mathematics, Peking University, Beijing, China

Pan, Chengdong 潘承洞
Shangdong University, Jinan, China

Pei, Dingyi 裴定一
Institute of Applied Mathematics, Academia Sinica, Beijing, China

Sun, Qi 孙琦
Institute of Mathematics, Sichuan University, Chengdu, China

Tsang, Kaiman 曾启文
The Department of Mathematics, Hong Kong University, Hong Kong

Wang, Yuan 王元
Institute of Mathematics, Academia Sinica, Beijing, China

Xie, Shenggang 谢盛刚
The Department of Mathematics, University of Science and Technology of China, Hefei, China

Xu, Guangshan 徐广善
Institute of Mathematics, Academia Sinica, Beijing, China

Ye, Yangbo 叶扬波
The Department of Applied Mathematics, Qing Hua University, Beijing, China

Contemporary Mathematics
Volume **77**, 1988

ANALYTIC NUMBER THEORY IN CHINA I

Chen Jingrun and Pan Chengbiao

The circle method and the methods of estimating the exponential sums are the most important methods in analytic number theory. By these methods a lot of important results have been obtained for many famous problems of number theory. In this paper, we'll give a survey of this kind of work made by Chinese mathematicians, and these works will be introduced in six parts. Mention must be made of late Professor Hua Lookeng in the first place, he made many important contributions to the development of the modern analytic number theory. There has been a systematic introduction of Hua's work on number theory made by Wang [1],[2] and Halberstam [1]. His work has also been well collected in his famous monographs [9],[10],[11],[12], Hua and Wang's [1], and other famous monographs, such as Vinogradov [1] and Vaughan [1]. Hence, we'll not give any detailed introduction of Hua's work here.

Throughout this paper we use the following notations:

$p, p', p_1, p_2, \ldots$	Prime numbers
$e(\theta)$	$\exp(2\pi i\theta)$
$C(\lambda,\ldots), C_1(\lambda,\ldots),\ldots$	Positive constants depending on parameters $\lambda,\ldots$

I. Estimation of exponential sums

[1] Complete exponential sums. Let q, k be positive integers, and

$$f(x) = a_k x^k + \ldots + a_1 x,$$

where a_j's are integers and $(a_k,\ldots,a_1,q) = 1$. Then the sum

$$S(q,f(x)) = \sum_{x=1}^{q} e\left(\frac{f(x)}{q}\right) \tag{1}$$

is called a complete exponential sum. The problem of estimating $S(q,f(x))$ has a long history, and it has very important applications in the theory of numbers. Gauss was the first to consider the simple case $f(x) = a_2x^2$, and $S(q,a_2x^2)$ is known as Gauss sum. He proved that

$$(2) \qquad |S(q,a_2x^2)| = \begin{cases} \sqrt{q}, & q \equiv 1 \pmod 2, \\ 0, & q \equiv 2 \pmod 4, \\ \sqrt{2q}, & q \equiv 0 \pmod 4. \end{cases}$$

In 1932, Mordell proved

$$(3) \qquad |S(p,f(x))| \le C_1(k)p^{1-(1/k)}$$

for prime p. For the general case, the problem was settled in 1940 by Hua [3] in a characteristically elegant way. He proved that for any $\varepsilon > 0$

$$(4) \qquad |S(q,f(x))| \le C_2(k,\varepsilon)q^{1-(1/k)+\varepsilon}.$$

Apart from the ε, this result is essentially best possible. We know from the famous work of A. Weil that the estimate (3) can be improved to

$$(5) \qquad |S(p,f(x))| \le (k-1)\sqrt{p}.$$

Using (5), the ε in Hua's estimate (4) can be omitted easily, and we get

$$(6) \qquad |S(q,f(x))| \le C_2(k)q^{1-(1/k)}.$$

Also using (5), Hua [7] proved that for any $\varepsilon > 0$ and $(a_k,q) = 1$,

$$(7) \qquad |S(q,a_kx^k+a_1x)| \le C_3(k,\varepsilon)(q,a_1)q^{(1/2)+\varepsilon}.$$

The evaluation of $C_2(k)$ was studied by several mathematicians. For example, Nechaev [1], Chen Jingrun [2], Nachaev [2], Chen Jungrun [14], Stechkin [1], Lu Minggao [5], Qi Minggao and Ding Ping [1,2], and Lu Minggao [6] proved $C_2(k) \le e^{2^k}$, $e^{3k^2+6/k}$, $e^{5k^2/\log k}$, $e^{6.1k}$ and $e^{k+O(k/\log k)}$, $e^{k+O(k/\log k)}$, $e^{2.33k}$, e^{2k}, and $e^{1.85k}$, respectively. For other refinements, see Loxton and Smith [1], Nachaev and Tupunov [1], Loxton and Vaughan [1], and Zhang Mingyao and Hong Yi [1].

For some other results on complete exponential sums, see Hua [6], Hua and Min [1], and Min [1].

In addition Yue Minyi [1] studied the following exponential sum

$$(8) \qquad \delta(n,\ell_1,\ell_2) = \sum_{\ell m=n} e\left(-\frac{\ell_1}{q_1}\ell+\frac{\ell_2}{q_2}m\right),$$

and obtained some applications for circle problem and divisor problem.

[2] Weyl exponential sums. Let $\alpha_k, \dots, \alpha_1$ be real numbers,

$$\phi(x) = \alpha_k x^k + \dots + \alpha_1 x,$$

and

$$T(P, \phi(x)) = \sum_{x=1}^{P} e(\phi(x)).$$

In 1938, using Weyl's method of estimating Weyl exponential sums, Hua [2] proved a mean value theorem for $T(P, \alpha_k x^k)$: For any $\varepsilon > 0$, we have

$$\int_0^1 |T(P, \alpha_k x^k)|^{2^\ell} d\alpha_k \ll P^{2^\ell - \ell + \varepsilon}, \quad 1 \le \ell \le k, \tag{9}$$

which is known as Hua's inequality.

In 1934, I.M. Vinogradov found a new powerful method for estimating Weyl exponential sums. Hua [5] improved and simplified Vinogradov's method by pointing out that the essence of this method is the following integral mean value theorem: For integers $k \ge 2$, $\ell \ge 0$, and $n \ge k(k+1)/4 + \ell k$, we have

$$\int_0^1 \dots \int_0^1 |T(P, \phi(x))|^{2n} d\alpha_1 \dots d\alpha_k \le (7n)^{4n\ell} P^{2n-(k^2+k)/2+\delta} (\log P)^{2\ell}, \tag{10}$$

where $\delta = (1/2)k(k+1)(1-(1/k))^\ell$. The estimate (10) is called Vinogradov-Hua mean value theorem, and has been improved further. In all current monographs on analytic number theory, Vinogradov's method is stated according to Hua's formulation. (See, for example, Vinogradov [1], Karacuba [1], and Vaughan [1].)

[3] Exponential sums with prime variable. In 1977, by using the complex integral method and some very elementary results on Dirichlet characters and L-functions, Pan Chengbiao [1] gave a much simpler analytic proof for the following estimation: For $(q,h) = 1$, $\lambda \ge 42$, and $(\log x)^\lambda \le q \le x(\log x)^{-\lambda}$ we have

$$\sum_{p \le x} e\left(\frac{h}{q} p\right) \ll x(\log x)^{-3}.$$

Yue [2] and Chen [10] obtained some earlier results for this kind of exponential sum. In 1984, Chen Jingrun [16] proved that for $x \ge 3$, $H = \exp(\sqrt{\log x}/2)$, and real numbers α satisfying

$$|\alpha - a/q| \le q^{-2} \quad (a,q) = 1,\ q \ge 1, \tag{12}$$

we have

$$\sum_{p \le x} e(\alpha p) \le 1.2x\,(\log x)^{0.75} \log\log x \left[\left(\frac{5}{q} + \frac{q \log q}{x}\right)^{1/2} + \frac{(\log q)^{1/2}}{H}\right]. \tag{13}$$

By Vaughan's method, Shan Zun [3] obtained independently the following result: Let $k > 1$ be an integer and α satisfy (12), then for any $\varepsilon > 0$,

$$\sum_{p \le x} e(\alpha p^k) \ll x^{1+\varepsilon}(q^{-1} + x^{-1/2} + x^{-k}q)^{4^{1-k}}. \tag{14}$$

A similar but more general result was obtained by Harman [1].

[4] By using exponential sum methods, Yue Minyi [1] and Jia Chaohua [1] studied some problems on square-free numbers. Let

$$\Delta(x) = \sum_{n \le x} \mu^2(n) - \frac{6}{\pi^2}x.$$

By improving the upper bounds for "type I" exponential sums, Jia [1] proved that $\Delta(x) \ll x^{7/22+\varepsilon}$ under the Riemann Hypothesis. The same result was obtained by Baker and Pintz [1] using a different technique. Recently, Jia has announced that he improved $7/22$ to some constant $< 6/19$.

II. Waring's problem

In 1770, E. Waring conjectured essentially that for any integer $k \ge 2$, there exists an integer $s = s(k)$ depending on k only such that every positive integer N can be expressed in the form

$$N = x_1^k + \dots + x_s^k, \tag{15}$$

where x_j's are non-negative integers. This conjecture was proved by Hilbert in 1909 using a very complicated method. After Hilbert, three problems were proposed for studying Waring's problem further, namely:

(i) Determine $g(k)$, the least value of s for which the Diophantine equation (15) is solvable for all N;

(ii) Determine $G(k)$, the least value of s for which the Diophantine equation (15) is solvable for all sufficiently large N;

(iii) Determine $\bar{G}(k)$, the least value of s for which an <u>asymptotic formula</u> for $r_s^{(k)}(N)$ can be found, where $r_s^{(k)}(N)$ is the number of the solutions of the Diophantine equation (15).

In the 20's of this century, Hardy, Ramanujan and Littlewood created and developed systematically a very powerful new analytic method - the so-called circle method - in additive number theory. Later, Vinogradov improved this method further. For Waring's problem, the circle method can be described as follows. It is clear that

$$r_s^{(k)}(N) = \int_0^1 T^s(P,\alpha x^k)e(-N\alpha)d\alpha, \tag{16}$$

where $P = [N^{1/k}]$. Then, roughly speaking, the interval $[0,1]$ is separated into two parts: E_1, consisting of short intervals centered at rational points a/q, $(a,q) = 1$, with small denominators q, and E_2, the rest of the unit interval. Thus we have

$$r_s^{(k)}(N) = \int_{E_1} + \int_{E_2} = r_{s1}^{(k)}(N) + r_{s2}^{(k)}(N). \tag{17}$$

Hardy and Littlewood proved that for $s \geq 2k+1$ we have

$$r_{s1}^{(k)}(N) = \Gamma^s(1+\frac{1}{k})\Gamma^{-1}(\frac{s}{k})\mathfrak{S}(N)N^{s/k-1} + o(N^{s/k-1}), \tag{18}$$

where singular series

$$\mathfrak{S}(N) = \sum_{q=1}^{\infty} \sum_{\substack{a=1 \\ (a,q)=1}}^{q} (\frac{1}{q}S(q,ax^k))^s e(-\frac{aN}{q}). \tag{19}$$

Hua [8] improved this result to $s \geq \max(5,k+1)$ in 1957. Hua's result is the best possible, and the proof is based on his estimate (7). Thus, the investigation of $r_s^{(k)}(N)$ is reduced to studying $\mathfrak{S}(N)$ and $r_{s2}^{(k)}(N)$; that is, to studying the complete exponential sum $S(q,ax^k)$ and the Weyl's sum $T(P,\alpha x^k)$. By Weyl's estimation for $T(P,\phi(x))$, Hardy and Littlewood proved that for $s \geq (k-2)2^{k-1}+5$, we have

$$r_{s2}^{(k)}(N) = o(N^{s/k-1}), \tag{20}$$

and

(21) $$r_s^{(k)}(N) \sim \Gamma^s(1+\frac{1}{k})\Gamma^{-1}(\frac{s}{k})\mathfrak{S}(N)N^{s/k-1}$$

Hence

(22) $$\bar{G}(k) \le (k-2)2^{k-1}+5.$$

In 1938, by his famous inequality (9), Hua [2] improved Hardy and Littlewood's result to $s \ge 2^k+1$; that is, (20) and (21) hold for $s \ge 2^k+1$, and

(23) $$\bar{G}(k) \le 2^k+1.$$

Recently, Vaughan [2], [4] has improved Hua's inequality (9), and proved that (20) and (21) hold for $s \ge 2^k$, $k \ge 3$; that is

(24) $$\bar{G}(k) \le 2^k, \quad k \ge 3.$$

By his method for estimating Weyl's sums, Vinogradov obtained a better result for large k:

(25). $$\bar{G}(k) \le 2k^2(2 \log k + \log\log k + 2.5), \; k \ge 11.$$

It should be pointed out that we can prove $\mathfrak{S}(N) > 0$ in <u>all</u> cases mentioned from (21) to (25).

Vinogradov also developed a method to treat $G(k)$, and proved

(26) $$G(k) \le \begin{cases} 3k \log k + 11\,k, \\ 2k \log k + 4k \operatorname{loglog} k + 2k \operatorname{logloglog} k + 13\,k, & k \ge 170{,}000. \end{cases}$$

Tong Kwangchang [1] and Chen Jingrun [1] obtained further results for $G(k)$.

For $g(k)$, Chen Jingrun [9], [13] proved that $g(5) = 37$ and $19 \le g(4) \le 27$. Recently, Deshouillers [1], [2] has announced that Balasubramaniall, Dress and himself proved that $g(4) = 19$.

For the history of Waring's problem and some new results, see Vaughan [1], [2], [3], [4], and Thanigasalam [2].

There are many ways to generalize Waring's problem, such as polynomial Waring problem, Waring-Goldbach problem, and so on. In China, Yang Wuzhi was the first one to study the cubic polynomial Waring problem. Hua made very important contributions to these problems, and his work can be found in his famous monographs [9], [11]. In addition, Chen Jingrun [2] also obtained some results for the polynomial Waring problem.

Recently, Lu Minggao [2], [3], [4], Lu and Shan Zun [1], and Shan Zun [1], [2], [4] have studied the Waring Goldbach problem and obtained some new results. Lu and Shan [1] proved that for any given positive integer k, almost all even positive integers N can be expressed by

$$N = p_1^2 + p_2^3 + p_3^5 + p_4^k. \tag{27}$$

Independent of Thanigasalam [1], Lu [17] proved that every sufficiently large integer N can be represented in the form

$$N = \sum_{j=1}^{22} x_j^{j+1}, \tag{28}$$

where x_j's are positive integers. Shan [1] proved that every sufficiently large odd integer N can be represented in the form

$$N = \sum_{j=1}^{23} p_j^{j+1} \quad \text{(with prime } p_j\text{)}. \tag{29}$$

Recently, Lu Minggao [7,(II)] has improved 22 to 17 in (28), and he [7,(III)] has established a lower bound for the correct asymptotic size for $R(n)$, the number of representations of n as the sum of one square and five cubes of natural numbers.

III. Goldbach's problem

In 1937, I.M. Vinogradov created an ingenious method for estimating exponential sums with prime variable and obtained the estimate (11) for $\lambda \geq \lambda_0$, and then using this result and the circle method, he proved that every sufficiently large odd integer is a sum of three primes. This result is known as the Goldbach-Vinogradov theorem, or three primes theorem. By the use of Vinogradov's method, several mathematicians, Hua among them, proved that almost all even integers can be represented as sums of two primes. (Hua [1] established more generally that almost all even integers can be represented as sums of a prime and a k-th power of a prime, where k is any given positive integer.) It was shown that, for any given $A > 0$,

$$E(x) \ll x(\log x)^{-A}, \tag{30}$$

where $E(x)$ denotes the number of the even integers $\leq x$ which are not sums of two primes. In 1975, Montgomery and Vaughan [1] proved that

$$E(x) \ll x^{1-\delta}, \tag{31}$$

where δ is some positive constant. Chen Jingrun and Pan Chengdong [1],

Chen [15], and Chen and Liu Jianmin [1] proved $\delta \geq 0.011$, 0.04, and 0.05, respectively. Lou Shiutou and Yao Qi [1], and Yao Qi [1] extended the estimate (31) to the case of small interval. Haselgrove, Pan Chengdong [1], and Chen Jingrun [12] discussed the following problem: Every sufficiently large odd integer N is a sum of three almost equal primes, that is,

$$N = p_1 + p_2 + p_3, \quad p_j \sim N/3, \quad j = 1,2,3.$$

In addition, Wu Fang [1], Lu Minggao and Chen Wende [1], and Lu Minggao [1] investigated systems of linear equations with prime variables.

Recently, Hua [13] and Pan Chengdong [2], [3], gave some interesting discussion about the Goldbach conjecture - every sufficiently large even integer is a sum of two primes.

IV. Estimation of the order of $\zeta(1/2+it)$

It is well known that $\zeta(1/2+it) \ll |t|^{1/4}$. Let θ be the lower bound of ξ such that

$$\zeta(1/2+it) \ll |t|^{\xi}.$$

Lindelöf conjectured that $\theta = 0$. This is a famous unsolved problem. Using the van der Corput's method of estimating the Weyl's sums, van der Corput and Koksma proved $\theta \leq 1/6$. By refining van der Corput's method, Titchmarsh, Min Sihe [2], and Yin Wenlin [1] obtained $\theta \leq 19/116$, $15/92$, and $13/80$ respectively. Later, Yin [4], Chen Jingrun [11] and Haneke [1] proved $\theta \leq 6/37$, and Kolesnik [1], [3], [4] proved $\theta \leq 173/1067$, $35/216$, and $139/858$.

Recently, Bombieri and Iwaniec [1], [2] developed a remarkable new method for bounding Weyl's sums, and obtained $\theta \leq 9/56$.

In addition, Hua and Wu Fang [1] investigated the problem of the order of $\zeta(\sigma+it)$ near $\sigma = 1$ by using Vinogradov's method for bounding Weyl's sums.

V. Gauss circle problem and sphere problem

Let $A_2(x)$ denote the number of the lattice points in the circle $u^2 + v^2 \leq x$. In the early 1800's, Gauss first proved that

$$A_2(x) = \pi x + O(\sqrt{x}).$$

To find α_2, the lower bound of all ξ such that

$$A_2(x) = \pi x + O(x^{\xi}),$$

is called the Gauss circle problem. In 1906, Sierpinski obtained $\alpha_2 \le 1/3$. By using his method for bounding Weyl's sum, van der Corput proved $\alpha_2 \le 37/112$. van der Corput's method and result were improved by several mathematicians. Until 1963, the best record was $\alpha_2 \le 13/40$, which was proved by Hua [4] in 1942. In 1963, Chen Jingrun [5] and Yin Wenlin [4] improved Hua's result to $\alpha_2 \le 12/37$. The latest record, $\alpha_2 \le 193/429$, is due to Nowak [1].

Let $A_3(x)$ denote the number of lattice points in the sphere $u^2+v^2+w^2 \le x$. To find α_3, the lower bound of all ξ such that

$$A_3(x) = \frac{4}{3}\pi x^{3/2} + O(x^{\xi})$$

is known as the sphere problem. In 1963, I.M. Vinogradov [1] and Chen Jingrun [6] proved $\alpha_3 \le 2/3$ independently.

Wu Fang [2] investigated the lattice point problem for an ellipse and found that

$$\sum_{au^2+2buv+cv^2\le x} 1 = \frac{\pi x}{\sqrt{ac-b^2}} + O(x^{13/40+\varepsilon}),$$

where a, b, c are real numbers, $a > 0$, and $ac - b^2 > 0$.

VI. Divisor problem

Let $d_k(n)$ be the number of ways representing n as a product of k factors, and $D_k(x) = \sum_{n\le x} d_k(n)$. It is well known that there exists a polynomial P_k of degree $k-1$ such that

$$\begin{cases} D_k(x) = xP_k(\log x) + \Delta_k(x), \\ \Delta_k(x) \ll x^{1-1/k}\log^{k-2} x, \end{cases} \qquad k = 2, 3, \ldots$$

One wants to find β_k, the lower bound of all ξ such that

$$\Delta_k(x) \ll x^{\xi}.$$

For $k = 2$, this is the famous Dirichlet divisor problem. In 1849, Dirichlet first proved that $P_2(y) = y+2\gamma-1$ and $\beta_2 \le 1/2$, where γ is Euler constant. In 1904, Voronoi proved $\beta_2 \le 1/3$. Using his method for estimating the Weyl's sums, van der Corput obtained $\beta_2 \le 33/100$ and $27/82$. By a refinement of van der Corput's method (which was proposed by Min Sihe for estimating the order of $\zeta(1/2+it)$), Chich Chungtao [1] proved $\beta_2 \le 15/46$ in 1950, and the same result was obtained by Richert in 1953 independently. Yin Wenlin [1], [4] obtained $\beta_2 \le 13/40$, $12/37$. In current record, $\beta_2 \le 139/429$, is due to Kolesnik [4].

For $k = 3$, Yue Minyi [4], Yin Wenlin [2], Yin [3], Yue Minyi and Wu Fang [1], Chen Jingrun [7], [8], Yin Wenlin [5], and Yin Wenlin and Li Zhongfu [1] proved

$$\beta_3 \le 14/29,\ 25/52,\ 10/21,\ 8/17,\ 5/11,\ 34/75, \text{ and } 127/282.$$

The best result, $\beta_3 \le 43/96$ is due to Kolesnik [2].

It should also be pointed out that, in principle, the Gauss circle problem and the Dirichlet divisor problem can be treated by the <u>same</u> methods used for estimating the order of $\zeta(1/2+it)$, and corresponding results obtained. Hence, using Bombieri-Iwaniec's method for estimating $\zeta(1/2+it)$, it seems that one should be able to obtain the corresponding results for these two problems; that is, $\alpha_2 \le 9/28$ and $\beta_2 \le 9/28$, but these remain to be confirmed.

To conclude this article, we wish to express our sincere gratitude to Professor Shan Zun and Professor Lu Minggao for providing us with some useful references.

References

R.C. Baker, J. Pintz

[1] The distribution of square-free numbers, Acta Arith. 46 (1985), 73-79.

E. Bombieri, H. Iwaniec

[1] Some mean-value theorems for exponential sums, to appear.

[2] On the order of $\zeta(1/2+it)$, to appear.

Chen Jingrun

[1] On Waring's problem for n-th powers, Acta Math. Sin. 8 (1958), 253-257.

[2] On the representation of natural number as a sum of terms of the form $x(x+1)\dots(x+k-1)/k!$, ibid, 9 (1959), 264-270.

[3] Improvement of asymptotic formulas for the number of lattice points in a region of the three dimensions, ibid, 12 (1962), 408-420.

[4] __________, Sci. Sin. 12 (1963), 151-161.

[5] The lattice points in a circle, Sci. Sin. 12 (1963), 633-649; Acta Math. Sin. 19 (1963), 299-313.

[6] Improvement on asymptotic formulas for the number of lattice points in a region of the three dimensions II, Sci. Sin. 12 (1963), 751-764.

[7] An improvement of asymptotic formulae for $\sum_{n\le x} d_3(n)$, Sci. Sin. 13 (1964), 1185-1188.

[8] On the divisor problem of $d_3(n)$, Acta Math. Sin. 14 (1964), 549-558; Sci. Sin. 14 (1965), 19-29.

[9] Waring's problem for $g(5) = 37$, Acta Math. Sin. 14 (1964), 715-734; Sci. Sin. 19 (1964), 1547-1568.

[10] Estimates for trigonometric sums, Acta Math. Sin. 14 (1964), 765-768.

[11] On the order of $\zeta(1/2+it)$, Acta Math. Sin. 15 (1965), 159-173; Sci. Sin 14 (1965), 522-538.

[12] On large odd number as sum of three almost equal primes, Sci. Sin. 14 (1965), 1113-1117.

[13] On an estimation of $g(4)$ of the Waring's probelm, Acta Math. Sin. 17 (1974), 131-142.

[14] On Professor Hua's estimate of exponential sums, Sci. Sin 20 (1977), 711-719.

[15] The exceptional set of Goldbach number (II), Sci. Sin. 26 (1983), 714-731.

[16] On the estimation of some trigonometrical sums and their application, Sci. Sin. 28 (1985), 449-458.

Chen Jingrun, Liu Jianmin

[1] The exceptional set of Goldbach number (III), Sci. Sin., to appear.

Chen Jingrun, Pan Chengdong

[1] The exceptional set of Goldbach number (I), Acta Sci. Nat. Univ. Shandong, 1979, no. 1, 1-27; Sci. Sin. 23 (1980), 416-430.

Chich Chungtao (Chih Tsung-tao)

[1] The Dirichlet's divisor problem, Sci. Rep. Nat. Tsing Hua Univ. A 5 (1950), 402-427.

J.-M. Deshouillers

[1] Problème de Waring pour les bicarrés: Le point en 1984, Groupe Étude Théor. Anal. Nombres, 1re/2e Années: 1984/1985, Exp. No. 33, 5 p. (1985).

[2] Problème de Waring pour les bicarrés, Sémin. Thèor,. Nombres, Univ. Bordeaux I 1984/1985, Exp. No. 14, 47 p. (1985).

H. Halbertstam

[1] Loo-Keng Hua: Obituary, Acta Arith., to appear.

W. Haneke

[1] Verschärfung der Abschätzung von $\zeta(1/2+it)$, Acta Arith. 8 (1962/63), 357-430.

G. Harman

[1] Trigonometric sums over primes I, Mathematika, 28 (1981), 249-254.

Hua Lookeng (Hua Loo-Keng)

[1] Some results in the additive prime-number theory, Quart. J. Math. Oxford 9 (1938), 68-80.

[2] On Waring's problem, Quart. J. Math. Oxford 9 (1938), 199-202.

[3] On an exponential sum, J. Chinese Math. Soc. 2 (1940), 301-312.

[4] The lattice points in a circle, Quart. J. Math. Oxford 13 (1942), 18-29.

[5] An improvement of Vinogradov's mean-value theorem and several applications, Quart J. Math. Oxford 20 (1949), 48-61.

[6] On exponential sums over an algebraic number field, Canadian J. Math. 3 (1951), 44-51.

[7] On exponential sums, Sci. Record (N.S) 1 (1957), 1-4.

[8] On the major arcs of Waring problem, ibid. 1 (1957), no. 3, 17-18.

[9] Additive Theory of Prime numbers, Sci. Press, Beijing, 1957.

[10] An Introduction to Number Theory, Sci. Press, Beijing, 1957.

[11] The Estimations of Exponential Sums and their Applications to Number Theory, Sci. Press, Beijing, 1963.

[12] Selected papers of L.-K. Hua, edited by H. Halberstam, Springer-Verlag, 1983.

[13] A direct attempt to Goldbach problem, Acta Math. Sin. New Ser., to appear.

Hua Lookeng, Min Sihe (Min Szu-Hoa)

[1] On a double exponential sum, Sci. Record 1 (1942), 23-25; Sci. Rep. Tsing Hua Univ. A 4 (1947), 484-518.

Hua Lookeng, Wang Yuan

[1] Application of Number Theory to Numerical Analysis, Sci. Press, Beijing, 1978.

Hua Lookeng, Wu Fang

[1] An improvement of Vinogradov's mean value theorem and some applications, Acta Math. Sin. 7 (1957), 574-589.

Jia Chaohua

[1] The distribution of square-free numbers, Acta Sci. Nat. Univ. Pekinensis, 1987, no. 3, 21-27.

A.A. Karacuba

[1] Principle of Analytic Number Theory, 2nd ed. Nauka, Moscow, 1983

G. Kolesnik

[1] On the estimation of certain trigonometric sums, Acta Arith. 25 (1973), 7-30.

[2] On the estimation of multiple exponential sums, in Recent Progress in Analytic Number Theory, Symp. Durham, 1979, Vol. 1 231-346, Academic Press, 1981.

[3] On the order of $\zeta(1/2+it)$ and $\Delta(R)$, Pacific J. Math. 98 (1982), 107-122.

[4] On the method of exponent pairs, Acta Arith. 45 (1985), 115-143.

Lou Shitou, Yao Qi

[1] The exceptional set of Goldbach numbers in a short interval, Acta Math. Sin. 24 (1981), 269-282.

J.H. Loxton, R.A. Smith

[1] On Hua's estimate for exponential sums, J. London Math. Soc. (2) 26 (1982), 15-20.

J.H. Loxton, R.C. Vaughan

[1] The estimation of complete exponential sums, Can. Math. Bull. 28 (1985), 440-454.

Lu Minggao

[1] A note on the systems of linear equations with prime variables, J. China Univ. Sci. Tech. 10 (1980), no. 4, 1-4.

[2] Improvement on a theorem of Roth's, ibid, supplement (1982), 13-18.

[3] On the problem concerning the sums of powers of natural numbers, ibid, supplement (1983), 16-31.

[4] Some problems in additive number theory (I), J. Math. Res. Exp. 3 (1984), 115-124.

[5] A note on the estimate of a complete rational trigonometric sum, Acta Math. Sin. 27 (1984), 817-823.

[6] The estimate of complete trigonometric sums, Sci. Sin. 28 (1985), 1-16.

[7] On the problem concerning the sums of powers of natural number (II), (III), to appear.

Lu Minggao, Chen Wende

[1] On the solution of systems of linear equations with prime variables, Acta Math. Sin. 15 (1965), 731-748.

Lu Minggao, Shan Zun

[1] A Waring Goldbach type problem, Kexue Tongbao, 27 (1982), 70-73.

Min Sihe (Min Szu Hoa)

[1] On systems of algebraic equations and certain multiple exponential sum, Quart. J. Math. Oxford 18 (1947), 133-142.

[2] On the order of $\zeta(1/2+it)$, Trans. Amer. Math. Soc. 65 (1949), 448-472.

H.L. Montgomery, R.C. Vaughan

[1] The exceptional set in Goldbach's problem, Acta Arith. 27 (1975), 353-370.

V.I. Nechaev

[1] On the representation of natural numbers as a sum of terms of the form $x(x+1)\ldots(x+n-1)/n!$, IAN SSSR Ser. Mat. 17 (1953), 485-498.

[2] An estimate of the complete rational trigonometric sum, Mat. Zametki 17 (1975), 839-849.

V.I. Nechaev, V.L. Topunov

[1] An estimate of the modulus of complete rational trigonometric sums of degrees, three and four, Proc Steklov Inst. Math. 1983, Issue 4, 135-140; Tr. Mat. Inst. Steklova 158 (1981), 125-129.

W.G. Nowak

[1] Zum Kreisproblem, Sitzungsber., Abt. II, Österr. Akad. Wiss., Math.-Naturwiss. Kl. 194 (1985), 265-271.

Pan Chengbiao

[1] A new proof of the three-prime theorem, Acta Math. Sin. 20 (1977), 206-211.

Pan Chengdong

[1] Some new results in the additive prime number theory, Acta Math. Sin. 9 (1959), 315-329.

[2] On the Goldbach problem, Acta Sci. Nat. Univ. Shandong, 1981, no. 1, 1-6.

[3] A new attempt on Goldbach conjecture, China Ann. Math. 3 (1982), 555-560.

Qi Minggao, Ding Ping

[1] Estimate of complete trigonometric sums, Kexue Tongbao, 29 (1984), 1567-1569.

[2] On Estimate of complete trigonometric sums, China Ann. Math. B 6 (1985), 110-120.

Shan Zun

[1] On a problem of sums of power of primes, J. China Univ. Sci. Tech 11 (1981), no. 4, 1-13.

[2] A question on sums of powers of integers, ibid. 12 (1982), no. 2, 1-11.

[3] On the estimate for $\sum e(\alpha p^k)$, ibid. Math. Issue, 1983, 154-155; J. Yangchow Normal College, 1983, no. 2, 7-13.

[4] Expressing n by $p_1^2 + p_2^4 + \dots + p_k^{2^k} + p_{k+1}^{2^k} + p_{k+2}^s$, J. Yangchow Normal College, 1983, no. 1, 1-3.

S.B. Stechkin

[1] Estimate of a complete rational trigonometric sum, Trudy Mat. Inst. Steklov 143 (1977), 188-207.

K. Thanigasalam

[1] On sums of powers and a related problem, Acta Arith. 36 (1980), 125-141; 42 (1983), 421.

[2] Improvement on Davenport's iterative method and new results in additive number theory, I, Acta Arith. 46 (1986), 1-31; II, ibid. 46 (1986), 91-112; III, ibid., to appear.

Tong Kwangchang

[1] On Waring's problem, China Adv. Math. 3 (1957), 602-607.

R.C. Vaughan

[1] The Hardy-Littlewood Method, Cambridge tracts 80, 1981.

[2] On Waring's problem for cubes, J. Reine Angew. Math. 365 (1986), 122-170.

[3] On Waring's problem for smaller exponents, Proc. London Math. Soc. (3) 52 (1986), 445-463.

[4] __________, II, Mathematika, 33 (1986), 6-22.

I.M. Vinogradov

[1] Selected Works, Springer-Verlag, 1984.

Wang Yuan

[1] Number theory, Hua [12], 1-5.

[2] Contributions of Professor Loo-Keng Hua to number theory, China Adv. Math. 15 (1986), 228-232.

Wu Fang

[1] On the solutions of the systems of linear equations with prime variables, Acta Math. Sin 7 (1957), 102-122.

[2] The lattice points in an ellipse, ibid. 13 (1963), 238-253.

Yao Qi

[1] The exceptional set of Goldbach numbers in a short interval, Acta Math. Sin. 25 (1982), 315-322.

Yin Wenlin

[1] On Dirichlet's divisor problem, Sci. Record (N.S.) 3 (1959), 6-8; acta Sci. Nat. Univ. Pekinensis 5 (1959), no. 2, 103-126.

[2] Piltz's divisor problem for $k = 3$, Sci. Record (N.S) 3 (1959), 169-173.

[3] On divisor problem for $d_3(n)$, Acta Sci. Nat. Univ. Pekinensis 5 (1959), no. 3, 193-196.

[4] The lattice-points in a circle, Sci. Sin. 11 (1962), 10-15; Acta Sci. (Nat.) Sichuan Univ. 1963.

[5] On divisor problem for $k = 3$, Acta Sci. (Nat.) Sichuan Univ. 1964, no. 1, 31-39.

Yin Wenlin, Li Zhongfu

[1] An improvement on the estimate for error term in the divisor problem for $d_3(n)$, Acta Math. Sin. 24 (1981), 865-878.

Yue Minyi (Yuh Ming-I)

[1] A note on the expressions and estimations of a kind of trigonometrical sum, Acta Math. Sin. 6 (1956), 35-54.

[2] Estimation of a trigonometric sum, ibid. 6 (1956), 105-114.

[3] On the differences between square-free numbers, Sci. Record (N.S.) 1 (1957), no. 3, 13-16.

[4] A divisor problem, Sci. Record (N.S.) 2 (1958), 326-328; Acta Math. Sin. 8 (1958), 496-506.

Yue Minyi, Wu Fang

[1] On divisor problems for $d_3(n)$, Acta Math. Sin. 12 (1962), 170-174; Sci. Sin. 11 (1962), 1055-1060.

Zhang Mingyao, Hong Yi

[1] On the maximum modulus of complete trigonometric sums, Acta Math. Sin., to appear.

Institute of Mathematics
Academia Sinica, Beijing

Department of Mathematics
Peking University, Beijing

Contemporary Mathematics
Volume **77**, 1988

ANALYTIC NUMBER THEORY IN CHINA II

Pan Chengdong   Pan Chengbiao
Xie Shenggang

By use of the circle method and the methods of estimating the exponential sums, Hua Lookeng and some other Chinese mathematicians obtained many important results in analytic number theory: these works are discussed in the paper "Analytic Number Theory in China I."

In this paper, we'll give a survey of the other work in analytic number theory done by Chinese mathematicians. These works will mainly relate to sieve methods and their applications; the theory of Riemann zeta function and Dirichlet L-functions and its applications; and to properties of some arithmetic functions. We'll introduce these works in three parts.

Throughout this paper we use the following notations:

$p, p_1, p_2, p', p_1', p_2'$	Prime numbers
$e(\theta)$	$e^{2\pi i\theta}$
$[x]$	Largest integer not exceeding x
$\|x\|$	$\mathrm{Min}\{x-[x], 1+[x]-x\}$
$\omega(n)$	Number of distinct prime divisors of integer n
$\Omega(n)$	Number of prime divisors of n, counted with multiplicity
$\mu(n)$	Möbius function: = 1, if $n = 1$; $(-1)^{\omega(n)}$, if n is square-free, 0, otherwise
$\Lambda(n)$	Mangoldt function: $= \log p$, if $n = p^k$, $k \geq 1$; 0, otherwise
$\phi(n)$	Euler totient function: the number of numbers $1,2,\ldots,n$ that are relative prime to n
$\pi(x)$	$\sum_{p\leq x} 1$
$\pi(x,q,\ell)$	$\sum_{x\geq p\equiv\ell(\mathrm{mod}\ q)} 1$
$\psi(x)$	$\sum_{n\leq x} \Lambda(n)$

$\psi(x,q,\ell)$	$\sum_{x\ge n\equiv\ell(\mathrm{mod}\ q)} \Lambda(n)$
γ	Euler constant, 0.577215...
$\chi(n) = \chi_q(n)$	Dirichlet character mod q
P_r	Almost prime, $\Omega(P_r) \le r$.

I. In this part we'll mainly discuss work on the following problems

(a) Propositions $\{r,s\}$ and $\{r,s\}_h$;

$\{r,s\}$: Every large even integer is a sum of P_r and P_s;

$\{r,s\}_h$: There are infinitely many integers n such that $\Omega(n) \le r$ and $\Omega(n-h) \le s$, h being a given integer.

(b) The upper bound for $D(N)$ and $Z_h(x)$.

$D(N)$: The number of solutions of the Diophantine equation $N = p_1+p_2$;

$Z_h(x)$: The number of solutions of the Diophantine equation $h = p-p'$, $p \le x$, h being a given even integer.

(c) The mean value theorem for distribution of primes in arithmetic progressions.

(d) Proposition $P(\lambda,r)$: For sufficiently large x, there exists an integer n satisfying $x-x^{\lambda} < n \le x$, and $\Omega(n) \le r$.

Before doing this, we'll briefly talk about some historical background concerning these works.

Let $\mathcal{A}$ be a finite sequence of integers, $\mathcal{B}$ an infinite set of primes, Θ a non-negative arithmetic function, and $Z \ge 2$. The sifting function with weight Θ is defined by

$$S(\mathcal{A},\mathcal{B},Z,\Theta) = \sum_{a\in\mathcal{A},\ (a,B(Z))=1} \Theta(a), \tag{1}$$

where

$$B(Z) = \sum_{Z>p\in\mathcal{B}} p. \tag{2}$$

The sieve method deals with estimates of the sifting function $S(\mathcal{A},\mathcal{B},Z,\Theta)$; here the sequence $\mathcal{A}$, the set $\mathcal{B}$ and the weight function Θ of course have to satisfy certain properties. Let $p(d)$ be a multiplicative function satisfying $0 \le \rho(p) < p$, $X > 2$, and

$$r_d(\Theta) = \mathcal{A}_d(\Theta) - \frac{\rho(d)}{\alpha}X, \tag{3}$$

where

(4) $$\mathcal{A}_d(\Theta) = \sum_{d|a\in\mathcal{A}} \Theta(a).$$

It is well known that V. Brun [1,2] was the first to give a pioneering great contribution to the sieve method in about 1920. He [2] constructed two sets $\mathcal{D}_1$ and $\mathcal{D}_2$ which depend on Z and some parameters, and satisfy some properties, two of which are that:

(i) if $d \in \mathcal{D}_i$ $(i = 1,2)$, then

(5) $$d|B(Z), \quad d < Z^{\nu_i},$$

where ν_i is a positive constant depending on the parameters;

(ii)

(6) $$X\sum_{d\in\mathcal{D}_2} \mu(d)\frac{\rho(d)}{d} + R_2(Z,\Theta) \le S(\mathcal{A},\mathcal{B},Z,\Theta) \le X\sum_{d\in\mathcal{D}_1} \mu(d)\frac{\rho(d)}{d} + R_1(Z,\Theta),$$

where

(7) $$R_i(Z,\Theta) = \sum_{d\in\mathcal{D}_i} \mu(d)r_d(\Theta), \quad i = 1,2.$$

Brun proved that there exist two positive constants λ_1,λ_2 depending on the parameters such that

(8) $$\lambda_2 XW(Z) + R_2(Z,\Theta) \le S(\mathcal{A},\mathcal{B},Z,\Theta) \le \lambda_1 XW(Z) + R_1(Z,\Theta),$$

where

(9) $$W(Z) = \prod_{Z>p\in\mathcal{B}} \left(1-\frac{\rho(p)}{p}\right).$$

For brevity, if $\Theta(n) = \Theta_0(n) \equiv 1$, we write $S(\mathcal{A},\mathcal{B},Z,)$, r_d, $\mathcal{A}_d$, $R_i(Z)$ for $S(\mathcal{A},\mathcal{B},Z,\Theta_0)$, $r_d(\Theta_0)$, $\mathcal{A}_d(\Theta_0)$, $R_i(Z,\Theta_0)$, respectively.

Let $\mathcal{P}$ be the set of all primes,

(10) $$\mathcal{A}^{(1)}(x,y) = \{a : x-y < a \le x\},$$

and

(11) $$\mathcal{A}^{(2)}(x,\ell) = \{a : a = n(\ell-n), \ 1 \le n \le x\},$$

ℓ being a given integer. Using his method, Brun [2] proved the following remarkable results.

Theorem 1. Propositions $\{q,q\}$ and $\{q,q\}_h$ are true. More precisely, for sufficiently large even integer N and sufficiently large x we have

$$S(\mathcal{A}^{(2)}(N);\mathcal{P},N^{1/10}) > 0.05\cdot 10^2\cdot 2e^{-2\gamma}C(N)N(\log N)^{-2},$$

and

$$S(\mathcal{A}^{(2)}(x,h);\mathcal{P},x^{1/10}) > 0.05\cdot 10^2\cdot 2e^{-2\gamma}C(h)x(\log x)^{-2},$$

where h is a given even integer, $\mathcal{A}^{(2)}(N) = \mathcal{A}^{(2)}(N,N)$, and

$$C(K) = \prod_{p>2}\left(1-\frac{1}{(p-1)^2}\right)\prod_{2<p|K}\frac{p-1}{p-2} \tag{12}$$

Theorem 2. For sufficiently large even integer N and sufficiently large x, we have

$$D(N) \le S(\mathcal{A}^{(2)}(N);\mathcal{P},N^{1/11}) + O(N^{1/11}) < 1.82\cdot 11^2\cdot 2e^{-2\gamma}C(N)N(\log N)^{-2}, \tag{13}$$

and

$$Z_h(x) \le S(\mathcal{A}^{(2)}(x,h);\mathcal{P},x^{1/11}) + O(x^{1/11}) < 1.82\cdot 11^2\cdot 2e^{-2\gamma}C(h)x(\log x)^{-2}.$$

Theorem 3. The proposition $P(1/2,11)$ is true. More precisely, for sufficiently large x we have

$$S(\mathcal{A}^{(1)}(x+x^{1/2}),\mathcal{P},x^{1/11}) > 0.3\cdot 11\cdot e^{-\gamma}x^{1/2}(\log x)^{-1}.$$

Brun's method and results were improved by many mathematicians. For brevity, in the following we'll discuss proposition $\{r,s\}$ and the upper bound for $D(N)$ only, since by the same methods we can obtain the similar results for $\{r,s\}_h$ and $Z_h(x)$ as we have seen in theorems 1 and 2.

An important contribution was made by A. A. Buchstab [1,2]. Using the combinatorial identity for sifting functions

$$S(\mathcal{A},\mathcal{B},Z,\Theta) = S(\mathcal{A},\mathcal{B},w,\Theta) - \sum_{w\le p|B(Z)} S(\mathcal{A}_p,\mathcal{B},p,\Theta)^{(*)}, \quad Z > w \ge 2, \tag{14}$$

which is now called Buchstab identity, Buchstab [1] obtained the following theorem on sifting functions.

Theorem 4. Let $f_0(\alpha)$ and $F_0(\alpha)$ $(2 \le \alpha \le 10)$ be two non-negative and

(*) Here $\mathcal{A}_d$ denotes the subsequence $\{a : a \in \mathcal{A},\ d|a\}$.

non-decreasing functions. If for large even integer N we have

$$f_0(\alpha)2e^{-2\gamma}C(N)N(\log N)^{-2} < S(\mathcal{A}(N),\mathcal{P},N^{1/\alpha}) < F_0(\alpha)2e^{-2\gamma}C(N)N(\log N)^{-2},$$

then for any β satisfying $2 \le \beta \le 10$, the inequality is still valid when $f_0(\alpha)$ and $F_0(\alpha)$ are replaced by the following functions $f_1(\alpha)$ and $F_1(\alpha)$ respectively:

$$f_1(\alpha) = \begin{cases} 0, & 2 \le \alpha < \tau, \\ f_0(\beta) - 2\displaystyle\int_{\alpha-1}^{\beta-1} F_0(t)\left(\frac{t+1}{t^2}\right)dt, & \tau \le \alpha < \beta, \\ f_0(\alpha), & \beta \le \alpha \le 10, \end{cases}$$

$$F_1(\alpha) = \begin{cases} F_0(\beta) - 2\displaystyle\int_{\alpha-1}^{\beta-1} f_0(t)\left(\frac{t+1}{t^2}\right)dt, & 2 \le \alpha < \beta, \\ F_0(\alpha), & \beta \le \alpha \le 10, \end{cases}$$

where τ is defined by

$$f_0(\beta) - 2\int_{\tau-1}^{\beta-1} F_0(t)\left(\frac{t+1}{t^2}\right)dt = 0.$$

Using Brun's method, Buchstab [1] proved that in theorem 4 we can take

$$(15)\qquad f_0(\alpha) = \begin{cases} 0.98\cdot 10^2, & \alpha = 10, \\ 0, \quad 2 \le \alpha < 10, \end{cases} \qquad F_0(\alpha) = 1.016\cdot 10^2, \quad 2 \le \alpha \le 10;$$

and later he [2] improved it to

$$(16)\qquad f_0(\alpha) = \begin{cases} 0.9998181\cdot 10^2, & \alpha = 10, \\ 0, \quad 2 \le \alpha < 10, \end{cases} \qquad F_0(\alpha) = 1.002073\cdot 10^2, \quad 2 \le \alpha \le 10.$$

Applying theorem 4 several times, he derived $f_1(6) > 0$ and $f_1(5) > 0$ from (15) and (16) respectively, hence proving propositions $\{5,5\}$ an $\{4,4\}$ respectively. In addition, he improved the upper bound for $D(N)$ also. Obviously, Buchstab's method can be used for other sifting functions.

Another important refinement of sieve method was made by A. Selberg [2,3] in about 1950. Let $\xi \ge 2$ be a parameter

$$G_1(\xi,Z) = \sum_{\xi > \ell \mid B(Z)} g(\ell),$$

where $g(1) = 1$

$$g(\ell) = \frac{\rho(\ell)}{\ell} \prod_{p|\ell} (1-\frac{\rho(p)}{p})^{-1}.$$

Selberg [2] proved that there exists a set of real numbers λ_d satisfying

$$\lambda_1 = 1, \ \lambda_d = 0, \ d \geq \xi,$$

such that

(17) $$S(\mathcal{A},\mathcal{B},Z,\Theta) \leq X(G_1(\xi,Z))^{-1} + R(Z,\xi,\Theta),$$

where

(18) $$R(Z,\xi,\Theta) = \sum_{\alpha|B(Z)} \left[\sum_{\substack{d_1<\xi,\, d_2<\xi \\ [d_1,d_2]=d}} \lambda_{d_1}\lambda_{d_2} \right] r_d(\Theta).$$

By his method, Selberg [3] improved (13) to

$$D(N) < 2(8+\varepsilon)C(N)N(\log N)^{-2},$$

ε being an arbitrary positive number. He [3] also announced that proposition $\{2,3\}$ can be derived by his methods, but no proof has ever appeared.

In about 1954, P. Kuhn [2] devised the so-called weighted sieve method, which is an important contribution to the application of sieve methods. Using his method and Brun's sieve, Kuhn proved the proposition $\{r,s\}$, $r+s \geq 6$, $r > 1$, $s > 1$.

So far, all the results on the proposition $\{r,s\}$ mentioned above pertain to the case $r > 1$, $s > 1$, and the proofs of these results are elementary. However, it is very difficult to prove proposition $\{1,5\}$. The main difference between these two cases lies in estimating the error terms R_1, R_2 and R. In 1932, by Brun's sieve, T. Estermann [1] proved

<u>Theorem 5</u>. Under the GRH, the proposition $\{1,6\}$ is true.

To prove unconditional result $\{1,s\}$, one needs a new idea and method. The pioneering contribution to proposition $\{1,s\}$ was made by A. Rényi [1] in 1948. By using Brun's sieve, Linnik's large seive [1], and the theory of Dirichlet L-functions, Rényi proved

<u>Theorem 6</u>. There exists an absolute constant s_0 such that the proposition $\{1,s_0\}$ is true.

Rényi's creative work will be introduced in detail later.

Building on these famous works of Brun, Buchstab, Selberg, Kuhn, Linnik, and Rényi, Chinese mathematicians have obtained many important results in the theory of sieve methods and their applications since the middle of the 1950's. Now we'll introduce these results.

[1] <u>The proposition</u> $\{r,s\}$. Let $z \geq y \geq 2$, H an integer and $P_{(H)}(\mathcal{A},\mathcal{B},z,y)$ denote the number of the elements of the subsequence

$$\{a : a \in \mathcal{A},\ (a, \prod_{z>p\in\mathcal{B}} p) = 1,\ (H-a, \prod_{\substack{y>p\in\mathcal{B}\\ p\nmid H}} p) = 1\}.$$

Let $u \geq v \geq 2$, N a large even integer. Using Buchstab's identity, Wang Yuan [1] proved that

$$\begin{aligned}(19)\qquad P_{(N)}(\mathcal{A}^{(1)}(N);\mathcal{P},N^{1/v},N^{1/u}) &\geq S(\mathcal{A}^{(2)}(N);\mathcal{P},N^{1/u})\\ &\quad - \sum_{N^{1/u}\leq p<N^{1/v}} S(\mathcal{A}^{(2)}(\tfrac{N}{p},N');\mathcal{P},N^{1/u})\end{aligned}$$

where $N' = N'(p)$ satisfies

$$N'p \equiv N \pmod{P(N^{1/u})},\quad p \geq N^{1/u}.$$

Using Brun-Buchstab's method to estimate the lower bound of $S(\mathcal{A}^{(2)}(N);\mathcal{P},N^{1/5})$, and Selberg's method to estimate the upper bound of the sum on the right of (19) with $u = 5$ and $v = 4$, Wang [1] obtained

<u>Theorem 7</u>. The proposition $\{3,4\}$ is true. More precisely,

$$P_{(N)}(\mathcal{A}^{(1)}(N);\mathcal{P},N^{1/4},N^{1/5}) > 1.00083\cdot 2e^{-2\gamma}C(N)N(\log N)^{-2}.$$

Furthermore, by refining and developing Kuhn's idea of weighted sieve method, Wang [3] devised the following weighted sieve:

$$\begin{aligned}(20)\qquad T(\mathcal{A}^{(2)}(N);u,v,m) &\geq S(\mathcal{A}^{(2)}(N);\mathcal{P},N^{1/u})\\ &\quad - \frac{2}{m+1}\sum_{N^{1/u}\leq p<N^{1/v}} S(\mathcal{A}_p^{(1)}(N);\mathcal{P},N^{1/u}) + O(N^{1-1/u}),\end{aligned}$$

where $u > v \geq 2$, integer $m \leq u$, and

$$(21)\qquad T(\mathcal{A},u,v,m) = \{a : a \in S(\mathcal{A},\mathcal{P},N^{1/u}),\ \mu(|a|) \neq 0,\ \sum_{\substack{N^{1/u}\leq p<N^{1/v}\\ p\mid a}} 1 \leq m\}.$$

Choosing suitably the parameters u,v and m, using Brun-Buchstab's method to estimate the lower bound of $S(\mathcal{A}^{(2)}(N);\mathcal{P},N^{1/u})$, and using Selberg's method to estimate the upper bound of the sum on the right of (20), Wang [3,5,8] proved the following theorem:

Theorem 8. We have

(a)

$$T(\mathcal{A}^{(2)}(N),6,3,2) > 0.33\cdot 2e^{-2\gamma}C(N)N(\log N)^{-2},$$

and hence the proposition {3,3} is true;

(b)

$$T(\mathcal{A}^{(2)}(N),8,2,3) > 0.56\cdot 2\cdot e^{-2\gamma}C(N)N(\log N)^{-2},$$

and hence the proposition {r,s}, $r+s \ge 5$, is true;

(c)

$$T(\mathcal{A}^{(2)}(N),8,6/7,2) > 0.43\cdot 2e^{-2\gamma}C(N)N(\log N)^{-2},$$

and hence the proposition {2,3} is true.

[2] Conditional results on the proposition {1,s} and the upper bound of D(N). Let L be a positive integer and

(22) $$\mathcal{A}^{(3)}(L) = \{a : a = L-p,\ p \le L\}.$$

Let $u \ge 2$. The proposition $\{1,-[-u]-1\}$ will be derived from that for large even integer N there is

(23) $$S(\mathcal{A}^{(3)}(N);\mathcal{P},N^{1/u},\Theta) > \theta(1).$$

Hereafter, we take

(24) $$\Theta(a) = \Theta_0(a) \equiv 1$$

or

(25) $$\Theta(a) = \Theta_1(a) = \Lambda(N-a)\exp\left\{-\frac{N-a}{N}\log N\right\}.$$

Applying Brun's seive, it is easy to prove that there are two non-negative and non-decreasing functions $f_0^*(u)$ and $F_0^*(u)$ such that

(26) $$\begin{aligned} f_0^*(u)e^{-\gamma}C(N)N(\log N)^{-2} + R_2^*(N^{1/u},\Theta) &\le S(\mathcal{A}^{(3)}(N);\mathcal{P},N^{1/u},\Theta) \\ &\le F_0^*(u)e^{-\gamma}C(N)N(\log N)^{-2} + R_1^*(N^{1/u},\Theta), \end{aligned}$$

where

(27) $$R_i^*(N^{1/u},\Theta) = \sum_{d|P_N(N^{1/u}),\, d\in\mathcal{D}_i} \mu(d) r_d^*(N,\Theta), \quad i = 1,2,$$

(28) $$r_d^*(N,\Theta) = \sum_{N\ge p\equiv N(d)} \Theta(N-p) - \frac{1}{\phi(d)} \sum_{p\le N} \Theta(N-p),$$

and

(29) $$P_k(z) = \prod_{z>p\nmid k} p.$$

Now, it is very difficult to estimate the error terms (27). It is well known that under the GRH we have

(30) $$\pi(x,d,\ell) - \frac{1}{\phi(d)}\pi(x) \ll x^{1/2}\log x, \quad (d,\ell) = 1.$$

Assuming (30) and using Brun's sieve, Estermann [1] proved that $f_0^*(7) > 0$ and $R_2^*(N^{1/7},\Theta_0) \ll N^{1-\delta}$, where δ is a positive constant, and hence theorem 5 follows.

By $H_0(\eta)$ ($\eta > 0$ we denote the following proposition: for any $B > 0$ and $\varepsilon > 0$ we have

(31) $$\sum_{d\le x^{\eta-\varepsilon}} \mu^2(d) \max_{(\ell,d)=1} |\pi(x;d,\ell) - \frac{1}{\phi(d)}\pi(x)| \ll \frac{x}{(\log x)^B};$$

and by $H_1(\eta)$ ($\eta > 0$) the following weaker proposition: for any $B > 0$ and $\varepsilon > 0$ we have

(32) $$\sum_{d\le x^{\eta-\varepsilon}} \mu^2(d) \max_{(\ell,d)=1} |\sum_{x\ge p\equiv\ell(d)} e^{-px/\log x} \log p - \frac{2}{\phi(d)}\frac{x}{\log x}| \ll \frac{x}{(\log x)^B}.$$

The problem of estimating the error terms appearing in Brun's sieve or Selberg's sieve can be reduced to proving the propositions $H_0(\eta)$ and $H_1(\eta)$. This kind of proposition is called a mean value theorem for the distribution of primes in arithmetic progressions. It is easy to see that the propositions $H_0(1/2)$ and $H_1(1/2)$ can be derived from (30). Using Buchstab's identity (14), Brun's sieve and Selberg's sieve, Wang Yuan [2] proved

Theorem 9. Assuming $H_0(1/2)$, the proposition {1,4} is true. More precisely,

$$S(\mathcal{A}^{(3)}(N);\mathcal{P},N^{1/5}) > 4.2e^{-\gamma}C(N)N(\log N)^{-2}.$$

Similar to (20), Wang devised the following weighted sieve:

$$T(\mathcal{A}^{(3)}(N),u,v,m) > S(\mathcal{A}^{(3)}(N);\mathcal{P},N^{1/u}) \tag{33}$$
$$- \frac{1}{m+1} \sum_{N^{1/u} \le p < N^{1/v}} S(\mathcal{A}_p^{(3)}(N);\mathcal{P},N^{1/u}) + O(N^{1-1/u}).$$

By suitable choice of the parameters and using Brun-Buchstab-Selberg's method, Wang [3,11] obtained the following results

Theorem 10. (a) Assuming $H_0(1/2)$, we have

$$T(\mathcal{A}^{(3)}(N),6,3,2) > 1.81e^{-\gamma}C(N)N(\log N)^{-2},$$

and hence $\{1,3\}$ is true; (b) assuming $H_0(1/3.237)$, we have

$$T(\mathcal{A}^{(3)}(N),5\cdot 3.237,\frac{20}{5-(3.237)^{-1}},1) > 0.01e^{-\gamma}C(N)N(\log N)^{-2},$$

and hence $\{1,4\}$ is true; (c) assuming $H_0(1/2.475)$, we have

$$T(\mathcal{A}^{(3)}(N),5\cdot 2.475,\frac{15}{5-(2.475)^{-1}},1) > 0.05e^{-\gamma}C(N)N(\log N)^{-2},$$

and hence $\{1,3\}$ is true. In addition, (d) assuming $H_0(1/2)$, we have for any $\varepsilon > 0$

$$D(N) < 2(4+\varepsilon)C(N)N(\log N)^{-2}. \tag{34}$$

Replacing $H_0(\eta)$ by $H_1(\eta)$ in theorems 9 and 10, we can obtain similar results.

[3] Proposition $\{1,s\}$, mean value theorem, and upper bound of $D(N)$. In 1948, Rényi [1] proved the following mean value theorem:

Theorem 11. There exist absolute constants u_1, u_2 such that

$$G_i^*(N^{1/u_i},\Theta_1) = \sum_{d\in\mathcal{D}_i,\, d|P_N(N^{1/u_i})} |r_d^*(N,\Theta_1)| \ll \frac{N}{(\log N)^3}, \quad i = 1,2.$$

It is easy to see that from theorem 11 and (26), theorem 6 follows at once. Rényi's proof of theorem 11 can be sketched as follows. By using Page's theorem (see Page, Proc. London Math. Soc., 39 (1935), 116-141), Siegel's theorem (see Siegel, Acta Arith., 1 (1936), 83-86), Brun-Titchmarsh theorem (see Titchmarsh, Rend. Cir. Mat. Palermo, 54 (1930), 414-429), and

(5), it is easy to prove that for any $\varepsilon > 0$ and $c_1 > 0$ we have

$$(35)\qquad G_i^*(N^{1/u},\Theta_1) = G_{i,1}^*(N^{1/u},\Theta_1) + O(N(\log N)^{-c_1+5} + N(\log N)^{-1/\varepsilon+3}),$$

where u is a positive number, and

$$(36)\qquad G_{i,1}^*(N^{1/u},\Theta_1) = \sum_d^{(i)} |r_d^*(N,\Theta_1)|,$$

the sum is over the range

$$(37)\qquad d \in \mathcal{D}_i,\ \exp((\log N)^{2/5}) < d < N^{\nu_i/u},\ \omega(d) < c_1 \log\log N.$$

Let p be the greatest prime factor of d, $\mu(d) \neq 0$, and $d = pq$. Then we have

$$(38)\qquad r_d^*(N,\Theta_1) = \frac{1}{\phi(p)} r_q^*(N,\Theta_1) + O(\textstyle\sum_1) + O(N^{1/2}),$$

where

$$(39)\qquad \sum\nolimits_1 = \frac{1}{\phi(d)} \sum_{\substack{\chi \bmod d \\ \chi=\chi_p\chi_q,\ \chi_p\neq\chi_p^0,\ \chi_q\neq\chi_q^0}} \Bigl| \sum_{\substack{\rho_\chi \\ |\tau|\le\log(N/\log N)}} \Gamma(\rho)\left(\frac{N}{\log N}\right)^{\rho} \Bigr|,$$

χ_k^0 denotes the principal character mod k, $\rho_\chi = \rho = \beta + i\tau$ is a non-trivial zero of $L(s,\chi)$, and the inner sum is over all the non-trivial zeros $\rho = \beta + i\tau$ of $L(s,\chi)$ satisfying $|\tau| \le \log(N/\log N)$. Roughly speaking, by improving Linnik's large sieve in various respects, and applying this method to treat the distribution of zeros of $L(s,\chi)$, Rényi proved that for almost all d, $\prod_{\chi \bmod d} L(s,\chi)$ has no zero in the domain

$$(40)\qquad 1 - c_2(\log d)^{-4/5} \le \sigma < 1,\quad |t| \le \log^3 d.$$

These d's are called non-exceptional moduli, and the others exceptional moduli. (See Rényi [1, Theorem 2] or Pan Chengdong [3, Theorem 3.1].) For non-exceptional moduli d satisfying (37), Rényi applied the zero-free region (40) to estimate the inner sum on the right of (39), and obtained from (38) and (39) that

$$(41)\qquad r_d^*(N,\Theta_1) = \frac{1}{\phi(p)} r_q^*(N,\Theta_1) + O(N^{1-c_2/(k_1+1)}),\quad 0 < k_1 < c_3 \log\log N,$$

c_3 being an absolute positive constant. On the other hand, by Brun-

Titchmarsh theorem we have

$$(42)\qquad \sum_{d \text{ exceptional}}^{(i)} |r_d^*(N,\Theta_1)| \ll N(\log N)^{-1/\varepsilon+3}.$$

Thus, using the property of the set $\mathcal{D}_i$, it follows from (36), (41) and (42) that

$$(43)\qquad G_{i,1}^*(N^{1/u},\Theta_1) = \sum_{p<N^{\nu_i/u}} \frac{1}{\phi(p)} \sum_{\substack{d \text{ non-exceptional} \\ d=pq}}^{(i)} |r_d^*(N,\Theta_1)|$$
$$+ O(N^{1-c_2/(k_1+1)+\frac{c_4}{u}k_1\nu_i h^{-k_1/2}}),$$

where h is the parameter in Brun's sieve. Applying the same method to deal with the inner sum on the right of (43), and repeating this procedure, we can finally get

$$(44)\qquad G_i^*(N^{1/u},\Theta_1) \ll N(\log N)^{-c_1+6} + N(\log N)^{-1/\varepsilon+4}$$
$$+ \log\log N \max_{k_1<c_3\log\log N} (N^{1-c_2/(k_1+1)+\frac{c_4}{u}k_1\nu_i h^{-k_1/2}}).$$

From this, theorem 11 follows at once.

Rényi didn't give the values of u_1, u_2 and s_0 explicitly, since it needs complicated computation for evaluating them. The shortcoming of Rényi's method is that the double sum in (39) is estimated by estimating its inner sum individually, and so u_1 and u_2 obtained by his method will be very small, and s_0 very large. Just noticing this shortcoming, Pan Chengdong applied his zero-density theorem of $L(s,\chi)$ [3, Theorem 2.1] and Rényi's theorem [1, Theorem 2] to estimate the double sum in (39) as a whole, and then improved (41) to

$$(45)\qquad r_d^*(N,\Theta_1) = \frac{1}{\phi(p)} r_q^*(N,\Theta_1) + O(N^{1-c_2/(k_1+1)}), \quad 0 < k_1 < c_3 \log\log N.$$

The same result can be obtained for $r_d^*(N,\Theta_1)$ replaced by

$$r_1(x,d,\ell) = \sum_{x\ge p\equiv \ell(d)} e^{-px/\log x} \log p - \frac{1}{\phi(d)} \frac{x}{\log x}.$$

From this and using the same argument, Pan Chengdong [3] proved

Theorem 12. Proposition $H_1(1/3)$ is true.

Furthermore, by use of Linnik's estimation [4] on the sixth moment of $L(s,\chi)$, Pan Chengdong [4] proved

Theorem 13. Proposiiton $H_1(3/.8)$ is true.

Using the simpler weighted sieve of Wang, Pan Chengdong derived proposition {1,5} [3] and {1,4} [4] from theorems 12 and 13 respectively. And Wang [11] derived the proposition {1,4} from theorem 12 and theorem 10(b). By the way, using these theorems the upper bound of $D(N)$ can be improved also.

Similar results on the mean value theorem and proposition {1,5} were obtained by Barban [1,2] independently.

The shortcoming of Pan and Barban's method is that the double sum in (32) is estimated by estimating its inner terms individually. In addition, it should be pointed out that the mean value theorem $H_0(\eta)$ is much more difficult to treat than $H_1(\eta)$, since the former needs to deal with the distribution of zeros of $L(s,\chi)$ with large imaginary part.

It is well known that Roth [1] and Bombieri [1] made important contributions on the large sieve. And, using the large sieve, Bombieri obtained the so-called large sieve type of zero-density theorem of $L(s,\chi)$ (see [1, Theorem 5]). A similar result was also obtained by A.I. Vinogradov (see [1, Theorem 1]). This type of theorem enables one to estimate the mean value in (31) as a whole, and then Bombieri [1] and Vinogradov [1] proved the proposition $H_0(1/2)$ independently in 1965. More precisely, Bombieri proved

Theorem 14. For any $B > 0$ there exists $A = A(B)$ such that

$$\sum_{d\le x^{1/2}\log^{-A}x} \max_{y\le x} \max_{(\ell,d)=1} \left|\pi(y;d,\ell) - \frac{1}{\phi(d)}\pi(y)\right| \ll \frac{x}{(\log x)^B}.$$

From $H_0(1/2)$, theorem 9 (c) and (d), proposition {1,3} and the upper bound (34) are derived unconditionally.

In 1965, by the use of Selberg's method, Jurkat and Richert obtained the best possible upper and lower estimations for the linear sifting functions (see [1, Theorem 4] or Rawsthorne [1]).

Let r be a positive integer, N a large even integer, and

$$\mathcal{A}^{(3)}_{(r)}(N) = \{a : a \in \mathcal{A}^{(3)}(N),\ \Omega(a) \le r\}. \tag{46}$$

In 1966, in order to prove {1,2}, Chen Jingrun [1] proposed a new weighted sieve as follows:

$$\mathcal{A}^{(3)}_{(2)}(N) \ge S(\mathcal{A}^{(3)}(N);\mathcal{P},N^{1/10}) - \tfrac{1}{2}\Omega_1 - \tfrac{1}{2}\Omega_2 + O(N^{9/10}), \tag{47}$$

where

$$\Omega_1 = \sum_{N^{1/10}\le p<N^{1/3}} S(\mathcal{A}_p^{(3)}(N);\mathcal{P},N^{1/10}), \tag{48}$$

$$\Omega_2 = \sum_{N^{1/10}\le p_1<N^{1/3}\le p_2<(N/p_1)^{1/2}} S(\mathcal{A}_{p_1p_2}^{(3)}(N);\mathcal{P},p_2). \tag{49}$$

And it was announced that by the weighted sieve (47) he proved

Theorem 15. The proposition {1,2} is true. More precisely

$$\mathcal{A}_{(2)}^{(3)}(N) > 0.67C(N)N(\log N)^{-2}. \tag{50}$$

Chen [3] published the proof of theorem 15 in 1973. The lower bound of $S(\mathcal{A}^{(3)}(N);\mathcal{P},N^{1/10})$ and the upper bound of Ω_1 can be obtained easily by the use of Jurkat-Richert's theorem and theorem 14, but we cannot apply the same way to estimate the upper bound of Ω_2. Chen noticed that for small $\varepsilon > 0$ there is

$$\Omega_2 \le S(\mathcal{L};\mathcal{P},N^{1/4-\varepsilon}) + O(N^{2/3}),$$

where

$$\mathcal{L} = \{\ell : \ell = N - p_1p_2p_3,\ N^{1/10} \le p_1 < N^{1/3} \le p_2 < (N/p_1)^{1/2},\quad p_2 < p_3 \le N/p_1p_2\}.$$

And then he used the original Selberg's upper bound sieve to estimate $S(\mathcal{L};\mathcal{P},N^{1/4-\varepsilon})$, but the error term is very complicated. By using theorem 14 and his skillful methods, Chen successfully estimated the error term, and obtained the upper bounds of $S(\mathcal{L};\mathcal{P},N^{1/4-\varepsilon})$ and Ω_2. From these, (50) follows.

Chen's argument is very complicated, and several simpler proofs were given by Halberstam and Richert [1], Pan, Ding and Wang [1], Halberstam [1], Ross [1], and Fujii [1]. It was pointed out by Pan and Ding [1,2] that the key point on the estimation of Ω_2 is the following mean value theorem.

Theorem 16. For any $B > 0$ and $0 < \alpha \le 1$, there exists $A = A(B)$ such that

$$\sum_{d\le x^{1/2}\log^{-A}x} \operatorname*{Max}_{y\le x} \max_{(\ell,d)=1} \Big| \sum_{\substack{a\le x^{1-\alpha}\\(a,d)=1}} \Big(\pi(y;a,d,\ell) - \frac{1}{\phi(d)}\pi\Big(\frac{y}{a}\Big)\Big)\Big| \ll \frac{x}{(\log x)^B},$$

where

$$\pi(y;a,d,\ell) = \sum_{y\geq ap\equiv \ell(d)} 1.$$

Clearly, Pan and Ding's theorem is a valuable generalization of theorem 14 (see Pan Chengdong [9]).

In 1978, Chen [6] improved the coefficient 0.67 in (50) to 0.81. Another remarkable result obtained by Chen [7] in 1978 is the following theorem.

Theorem 17. For sufficiently large even integers N, we have

$$D(N) < 7.834C(N)N(\log N)^{-2}. \tag{51}$$

For proving (51), Chen devised another new kind of weighted sieve. Although the key point of realizing his method still is theorem 16, his argument is very complicated. A simpler proof was given by Pan Chengbiao [4], but Pan's result is weaker than (51), that is, 7.834 is replaced by 7.928. The simplest case of Chen's weighted sieve used by Pan is as follows. By Buchstab's identity, we have

$$S(\mathcal{A}^{(3)}(N);\mathcal{P},N^{1/5}) \leq S(\mathcal{A}^{(3)}(N);\mathcal{P},N^{1/7}) - \tfrac{1}{2}\Omega_3 + \tfrac{1}{2}\Omega_4 + O(N^{6/7}), \tag{52}$$

where

$$\Omega_3 = \sum_{N^{1/7}\leq p_1<N^{1/5}} S(\mathcal{A}^{(3)}_{p_1}(N);\mathcal{P},N^{1/7}), \tag{53}$$

$$\Omega_4 = \sum_{N^{1/7}\leq p_2<p_3<p_1<N^{1/5}} S(\mathcal{A}^{(3)}_{p_1p_2p_3}(N);\mathcal{P}(p_2),p_3), \tag{54}$$

$$\mathcal{P}(K) = \{p : p\nmid K\}.$$

Using Jurkat-Richert theorem and theorem 14 we can easily get the upper bound of

$$S(\mathcal{A}^{(3)}(N);\mathcal{P},N^{1/7}) - \tfrac{1}{2}\Omega_3.$$

By the same way of estimating Ω_2, we can obtain the upper bound of Ω_4. From these estimations and

$$D(N) \leq S(\mathcal{A}^{(3)}(N);\mathcal{P},N^{1/5}) + O(N^{1/5}).$$

Pan's weaker result follows.

By the way, the upper bound of $Z_h(x)$ can be further improved by using a new refinement of Bombieri's theorem, but this method cannot be used to treat $D(N)$.

Using Chen's weighted sieve and Pan-Ding's mean value theorem, E.K.-S. Ng [1,2], Zhang Mingyao [3,4,5], and Shao Xiong [1] obtained some interesting results on the propositions $\{r,s\}$, $\{r,s\}_h$ and their generalizations. For example:

(i) Shao Xiong [1] proved that

$$\mathcal{A}_{(3)}^{(3)}(N) > 6.82C(N)N(\log N)^{-2},$$

which is an improvement of Ng's result [1]. Zhang Mingyao [4] proved that

$$\mathcal{A}_{(4)}^{(3)}(N) > 9.3153C(N)N(\log N)^{-2},$$

and

$$\{a = N - p_1p_2 : p_1p_2 < N, \ \Omega(a) = 2 \ \text{ or } \ 3\} > 5.3272C(N)N(\log N)^{-2},$$

and some other similar results.

(ii) Zhang Mingyao [3,5] proved the following theorem and some other results which are generalizations of Ng's results [2]. For any given integers $r \geq 1$ and $s \geq 1$, there exist positive constants $N(r,s)$ and $c(r,s)$ such that for any even integer $N \geq N(r,s)$, the number of the solutions for the Diophantine equations $N = p_1 \cdots p_r + p_1' \cdots p_s'$ or $p_1 \cdots p_r + p_1' \cdots p_{s+1}'$ is

$$\gg c(r,s)C(N)N(\log N)^{-2}.$$

Recently, Zhan Tao [4] has proved that for any given positive number A,

$$\sum_{q \leq x^{1/38.5}} \max_{(a,q)=1} \max_{h \leq H} \max_{x/2 < y \leq x} \left|\psi(y+h;q,a) - \psi(y;q,a) - \frac{h}{\phi(q)}\right| \ll \frac{H}{\log^A x},$$

provided $H = x^{\theta}$, $7/12 < \theta \leq 1$, which is an improvement of Perelli-Pintz-Salerno's result. In addition, Zhang Dexian [1,3] obtained an extension of Pan-Ding's mean value theorem for $\mu(n)$, and that of Barban's mean value theorem for $\mu(n)$.

[4] <u>Almost primes in short intervals</u>. In 1953, by using his weighted sieve, Kuhn [1] improved Brun's theorem 3, and obtained

<u>Theorem 18</u>. For any positive integer α, the proposition $P(1/\alpha, \alpha+\beta)$ is

true, provided β is the smallest integer satisfying the inequality

$$\log(b\alpha-\beta) \le 0.968(\beta+1),$$

especially, the proposition $P(1/2,4)$ is true.

Kuhn's method was further developed by Wang [4,9]; he devised the following weighted sieve

$$\begin{aligned} T^*(\mathcal{A}^{(1)}(x,x^{1/\alpha});u,v,m) \ge\ & S(\mathcal{A}^{(1)}(x,x^{\alpha});\mathcal{P},x^{1/u}) \\ & - \frac{1}{m+1} \sum_{x^{1/u}\le p<x^{1/v}} S(\mathcal{A}_p^{(1)}(x,x^{\alpha});\mathcal{P},x^{1/u}) \\ & + O(x^{1-1/u}+x^{1/v}), \end{aligned} \tag{55}$$

where α is a positive number, m a positive integer, $u > v > 1$, and

$$\begin{aligned} T^*(\mathcal{A}^{(1)}(x,x^{1/\alpha});u,v,m) = \{a : a \in \mathcal{A}^{(1)}(x,x^{1/\alpha}), (a,P(x^{1/u})) = 1, \\ \mu((a, \prod_{x^{1/u}\le p<x^{1/v}} p^2)) \ne 0,\ \Omega((a, \prod_{x^{1/u}\le p<x^{1/v}} p)) \le m\}. \end{aligned}$$

By suitable choice of the parameters α,u,v,m in (55), and using Brun-Buchstab-Selberg's method, Wang proved the following

<u>Theorem 19</u>. For any positive number α, the proposition $P(1/\alpha,\alpha+\beta)$ is true, provided β is the smallest integer satisfying

$$5.64527 + 3.65 \log \frac{5\alpha-\beta}{5+\beta} \le 4.8396(\beta+1).$$

Furthermore, the proposition $P(10/17,2), P(20/49,3)$, and $P(1/5,6)$ are true.

Wang's results were improved by Jurkat and Richert [1], and Richert [1], especially they proved the propositions $P(14/25,2)$, and $P(6/11,2)$. Similar results concerning the problem of almost primes representable by polynomials were also obtained by Kuhn, Wang, Jurkat and Richert. An important contribution on proposition $P(\lambda,r)$ and sieve theory was made by Chen Jingrun [4] in 1975, when he proved

<u>Theorem 20</u>. The proposition $P(1/2,2)$ is true.

Chen's argument is as follows. Let $\alpha > 1$, r a positive integer, and

$$M(x,\alpha,r) = \{a : x - x^{1/\alpha} < a \le x,\ \Omega(a) \le r\}.$$

He devised the following weighted sieve with logarithmic weight function:

(56)
$$\frac{18}{7}M(x;2,2) \geq S(\mathcal{A}^{(1)}(x,x^{1/2});\mathcal{P},x^{1/10}) - \sum_{x^{1/10}\leq p<x^{1/7}} S(\mathcal{A}_p^{(1)}(x,x^{1/\alpha});\mathcal{P},p)$$
$$-\frac{9}{7}\sum_{x^{1/7}\leq p<x^{9/20}} \left(1-\frac{20}{9}\,\frac{\log p}{\log x}\right)S(\mathcal{A}_p^{(1)}(x,x^{1/2});\mathcal{P},x^{1/7}).$$

As usual, the lower bound of $S(\mathcal{A}_p^{(1)}(x,x^{1/2});\mathcal{P},x^{1/10})$ can be easily estimated by Jurkat-Richert's theorem [1, Theorem 4]. In order to get good upper bounds for the two sums on the right of (56), Chen devised a new method to estimate the following type of sum

$$I(u_1,u_2,\beta) = \sum_{x^{u_1}\leq p\leq x^{u_2}} S(\mathcal{A}_p^{(1)}(x,x^{1/2});\mathcal{P},x^{\beta}),$$

where $0 < \beta < u_1 < u_2$. His method can be described as follows. At first, using Selberg's method he obtained

$$I(u_1,u_2,\beta) = \sum_{x^{u_1}\leq p\leq x^{u_2}} \sum_{a\in\mathcal{A}_p^{(1)}(x,x^{1/2})} \left[\sum_{d|(a,p(x^{\beta}))} \lambda_d\right]^2$$
$$\leq \frac{x^{1/2}}{\beta\log x} \sum_{x^{u_1}\leq p\leq x^{u_2}} \frac{1}{p} + R,$$

where λ_d's are suitably chosen and satisfy

$$\lambda_1 = 1,\ |\lambda_d| \leq 1, \ \text{ and } \ \lambda_d = 0 \ \text{ if } \ d \geq x^{\beta};$$

and

$$R = -\sum_{1\leq d\leq x^{2\beta}} f(d) \sum_{x^{u_1}\leq p\leq x^{u_2}} \left[\left\{\frac{x}{pd}\right\} - \left\{\frac{x-x^{1/2}}{pd}\right\}\right],$$

$$f(d) = \sum_{x^{2\beta}\geq d|P(x^{\beta})} \sum_{[d_1,d_2]=d} \lambda_{d_1}\lambda_{d_2}.$$

Secondly, using the expansion into Fourier series of $\{x/pd\} - \{(x-x^{1/2})/pd\}$, Chen proved that for any $0 < \Delta \leq 1$,

$$R \ll x^{u_2+2\beta}\Delta + \sum_{m=1}^{\infty} Z_m \sum_{x^{u_1}\le p\le x^{u_2}} \left\{ \Big|\sum_{1\le d\le x^{2\beta}} f(d)e\Big(\frac{mx}{pd}\Big)\Big| + \Big|\sum_{1\le d\le x^{2\beta}} f(d)e\Big(\frac{m(x-x^{1/2})}{pd}\Big)\Big| \right.$$

(57)

$$\left. + \Big|\sum_{1\le d\le x^{2\beta}} g(d)e\Big(\frac{mx}{pd}\Big)\Big| + \Big|\sum_{1\le d\le x^{2\beta}} g(d)e\Big(\frac{m(x-x^{1/2})}{pd}\Big)\Big| \right\},$$

where

$$g(d) = \sum_{x^{2\beta}\ge d\mid P(x^{\beta})} \sum_{[d_1,d_2]=d} 1,$$

and

$$Z_m = \begin{cases} m^{-1}, & m \le \Delta^{-1}, \\ (\Delta^2 m^3)^{-1}, & m > \Delta^{-1}. \end{cases}$$

Finally, he applied van der Corput's method to estimate the upper bounds of the exponential sums on the right of (57). Thus, a good upper estimation of $I(u_1,u_2,\beta)$ was obtained. By his method Chen proved

$$\frac{18}{7}M(x;2,2) > 0.14\,\frac{x^{1/2}}{(\log x)^2},$$

and then theorem 20 follows.

Four years later, Chen [8] improved his method and proved

Theorem 22. The proposition $P(0.477,2)$ is true.

Chen was the first one to apply the methods for estimating exponential sums to the estimation of the error term in sieve methods. This is a great contribution to the theory of sieve methods. His innovation inspired Iwaniec in the discovery of a powerful new version of the linear sieve (see Iwaniec [1], Halberstam, Heath-Brown, and Richert [1]).

Chen's result was improved by several mathematicians. Halberstam, Heath-Brown, and Richert [1], Iwaniec and Laborde [1], and Halberstam and Richert (to appear) improved 0.477 to 0.455, 0.45, and 0.4476 respectively.

[5] By sieve methods or by combining sieve methods and other methods, Chinese mathematicians also obtained some other results which we'll briefly talk about.

(a) Let $g(p)$ be the last positive primitive root mod p. By the method of Burgess, Wang Yuan [10] and Burgess [1] proved $g(p) \ll p^{1/4+\varepsilon}$ independently, where ε is an arbitrary positive number. In addition, under

GRH, Wang [10] proved $g(p) \ll m^6\log^2 p$, where $m = \omega(p-1)$. Recently, by Rosser-Iwaniec sieve, Lu Minggao [1] improved m^6 to $m^{4+\varepsilon}$.

Wang [9] and Pan Chengdong [5] also obtained some results on the estimation of the upper bound of the least positive non-residue of k-th degree modulo p.

(b) Let $N(s)$ be the maximal number of pairwise orthogonal Latin squares of order s. In 1966, Wang Yuan [13] proved that for sufficiently large s, $N(s) > s^{1/26}$, and in 1984, Lu Minggao [2] improved 1/26 to 10/143.

(c) Let k be a positive integer a_i, b_i $(i = 1,\dots,k)$ integers, and $F_k(n) = \prod_{i=1}^{k}(a_i n+b_i)$ have no fixed prime factor, and let r_k be the least positive integer such that there are infinitely many n satisfying $\Omega(F_k(n)) \le r_k$. In 1965, Xie Shenggang [1] proved $r_3 \le 15$; in 1983k he [2] proved that $r_4 \le 14$, $r_5 \le 18,\dots,r_{16} \le 79$, and that for $k \ge 17$, $r_k \le k\log\nu_k + k + (0.0139)k^{-1}$, where ν_k is a constant depending on k only and satisfying $\lim_{k\to\infty}\nu_k/k = 2.444\dots$. And Xie [3] proved $r_2 \le 3$ in 1983. In addition, let a, b, and m be positive integers satisfying $(a,b) = 1$, $m = 1$ or 2, and $a+b \equiv m \pmod 2$, Xie [6] discussed the equation $ap - bP_2 = m$.

(d) Goldbach-Schnirelman problem. In 1956, Yin Wenlin [1,2] proved that every sufficiently large integer is a sum of at most 18 primes. Vaughan [1] improved 18 to 6 in 1977. In 1982, Zhang Mingyao and Ding Ping [1], Zhang Mingyao [1] proved that every positive integer is a sum of at most 24 primes, and the best result 19 is due to Riesel and Vaughan [1].

(e) Wang Yuan [7] and Shao Pinzong [1-5] obtained some interesting results on the distribution of the ratios of some arithmetic functions. For example, Wang [7] and Shao [4] proved that for any given non-negative numbers $a_1,\dots,a_k$, and $\varepsilon > 0$, there is a prime p such that

$$\left|\frac{f(p+\ell+1)}{f(p+\ell)} - a_\ell\right| < \varepsilon, \quad 1 \le \ell \le k,$$

where $f(n)$ is $\phi(n)$, or $\omega(n)$ or $\Omega(n)$.

(f) Let $p(x,y)$ denote the greatest prime factor of $\prod_{x<n\le x+y} n$. by the use of sieve method and exponential sum method,. Graham [2] proved that for sufficiently large x, $p(x,x^{1/2}) > x^{0.66}$. In 1985, by improving the upper bounds of "type I" exponential sums, Jia Chaohua [1] improved 0.66 to $(23/48)^{1/2} - \varepsilon$, ε being any small positive number. By using Iwaniec's sieve method and many new devices of estimating exponential sums, R.C. Baker [1] proved that 0.66 can be replaced by 0.70. Recently, Jia Chaohua [3] has further improved 0.70 to 0.71 and 0.716 by combining Baker's and his arguments.

(g) Let $G(n)$ be the number of non-isomorphic groups of order n, and $F_k(x)$ the number of all the integers n satisfying $n \le x$, $G(N) = k$. Lu Minggao [5] proved that

$$F_2(x) = \frac{e^{-\gamma}x}{(\log_3 x)^2} + O\left(\frac{x(\log_4 x)^2}{(\log_3 x)^3}\right),$$

where $\log_{r+1} x = \log(\log_r x)$. Furthermore, Liu Hongquan [1] obtained that for every integer $a \ge 2$, we have

$$F_{2^a}(x) = \frac{e^{-\gamma}x}{a!\,(\log_3 x)^{a+1}} + O\left(\frac{x(\log_4 x)^{a+1}}{(\log_3 x)^{a+2}}\right).$$

In addition, Zhang Mingyao [2] proved the following results:

$$F_1(x) = \frac{e^{-\gamma}x}{\log_3 x} + O\left(\frac{x(\log_4 x)}{(\log_3 x)^2}\right),$$

and

$$\sum_{n\le x} \mu^2(n)\log G(n) = \left(\frac{6}{\pi^2}\sum_p \frac{\log p}{p^2-1}\right)x \log_2 x + O(x \log_3 x).$$

[6] Large sieve and its applications. Lu Minggao [1] proved the following result. Let N be a positive integer, $0 < \delta \le 1/2$, and $x_1, \dots, x_R$ any real numbers satisfying $\|x_r - x_s\| \ge \delta$ when $r \ne s$. Then if $N\delta \le 1/4$ we have

$$\sum_{r=1}^{R} \left| \sum_{n=M+1}^{M+N} a_n e(nx_r) \right|^2 < \delta^{-1}(1+22N^3\delta^3) \sum_{n=M+1}^{M+N} |a_n|^2,$$

a_n being complex numbers.

Zhang Dexian [2] obtained the following upper bound estimation. Let m be a positive integer $\varepsilon > 0$, $\delta > 0$ and $N \le Q\,(\log Q)^{\delta}$. Then we have

$$\sum_{p\le Q} \sum_{\substack{b=1 \\ p\nmid b}}^{p^m} \left| \sum_{n=M+1}^{M+N} a_n e\left(\frac{nb}{p^m}\right) \right|^2 \ll \frac{Q^{1+m}}{(\log Q)^{1-\varepsilon}} (\log Q)^{(\varepsilon+\delta)(1-1/m)} \sum_{n=M+1}^{M+N} |a_n|^2.$$

By using the arithmetic application of the large sieve, Shen Zun [1] proved the following theorem. Let $E_{a,k}(N)$ denote the number of natural numbers $n \le N$ for which equation

$$\sum_{i=0}^{k}\frac{1}{x_i} = \frac{a}{n}$$

is insoluble in positive integer x_i $(i = 0, 1, \dots k)$. Then

$$E_{a,k}(N) \ll N \exp\{-c(\log N)^{1-1/(k+1)}\}$$

II. In this part we'll introduce some work on the theory of $\zeta(s)$ and $L(s, \chi)$, and its applications.

[1] The least prime in an arithmetic progression. In 1944, Linnik [2] proved two theorems on the distribution of zeros of $L(s, \chi)$. Using these two theorems, Linnik derived the following famous result. Let $0 \le \ell \le q$, $(\ell, q) = 1$, and $p(q, \ell)$ the least prime in the arithmetic progression: $\ell + dq$, $d = 0, 1, 2, \dots$. Then there are constants L and q_0 such that

$$p(q, \ell) \le q^L, \quad q \ge q_0.$$

Linnik's argument is very complicated, and Rodosskii [1] simplified his proof. But they didn't give the constant L explicitly. By Linnik-Rodosskii method, Selberg sieve, and some analysis tricks from the theory of $L(s, \chi)$, Pan Chengdong [1] proved $L \le 5448$. In 1964, Chen Jingrun [1] improved it to $L \le 770$.

Using Turan's power-sum method, Linnik's argument was simplified greatly, and by this method, Jutila [1] proved $L \le 550$ in 1970. And then Chen [5] improved it to $L \le 168$.

In 1977, by the use of Halász's method, Selberg's devise of pseudo-characters, and an elegant asymptotic formula due to Graham, Jutila [2] obtained that $L \le 80$. Subsequently, Graham [1], Chen [9], and Wang Wei [1,2,3] improved the constant 80 to 20, 17, and 16 respectively. In addition, Chen [10] has announced that $L \le 15$, and now he has further obtained $L \le 13.5$.

[2] The distribution of zeros of $\zeta(s)$ and $L(s, \chi)$. Let T be a sufficiently large positive number, $1/2 \le \alpha \le 1$, $N(\alpha, T)$ the number of zeros of $\zeta(s)$ in the region $\alpha \le \sigma \le 1$, $0 \le t \le T$, and $N_0(T)$ the number of the zeros of $\zeta(s)$ on the line $\sigma = 1/2$, $0 \le t \le T$. Many mathematicians, such as Hardy, Littlewood and Selberg, studied the problem of the lower bound for $N_0(T)$. Min Szuhao proved that $N_0(T) > (60{,}000)^{-1}N(1/2, T)$. An important

contribution was made by Levinson [1,2,3] in 1974 and 1975; he proved $N_0(T) > 0.34N(1,T)$ and $> 0.3474N(1/2,T)$ respectively. In 1977 Pan Chengbiao [1] simplified Levinson's argument, and another simplification was given by Conrey and Ghosh [1]. In 1979 Lou Shituo [1], Lou and Yao Qi [1] proved $N_0(T) > 0.35N(1/2,T)$, and in 1982 Mo Guoduan [1] proved $N_0(T) > 0.3654N(1/2,T)$. The best result $N_0(T) > 0.3658$ is due to Conrey [1].

The estimation of the upper bound of $N(\alpha,T)$ is very important in analytic number theory. Many results have been obtained for this problem. In 1965, Pan Chengdong [7] obtained an estimation for $N(\alpha,T)$, and in 1985, Zhang Yitang [1] proved that

$$N(\alpha,T) \ll T^{A(\alpha)(1-\alpha)+\varepsilon}, \quad (13/17 < \alpha < 1)$$

where ε is any positive number,

$$A(\alpha) = \begin{cases} 3/2\alpha, & (4/5 \le \alpha < 1), \\ 3/(7\alpha-4), & (11/14 \le \alpha < 4/5), \\ 9/(7\alpha-1), & (41/53 \le \alpha < 11/14), \\ 7/(29\alpha-19), & (107/139 \le \alpha < 41/53), \\ 6/(5\alpha-1), & (13/17 \le \alpha < 107/139). \end{cases}$$

Let $N(\alpha,T,\chi)$ denote the number of zeros of $L(s,\chi)$ in the region $\alpha \le \sigma \le 1$, $|t| \le T$, χ^0 the principal character, and

$$N'(\alpha,T,q) = \sum_{\chi^0 \ne \chi \bmod q} N(\alpha,T,\chi).$$

Chen [11] proved that, let $q \ge 3$, $T \ge \mathrm{Max}(10^5 q^{-1}, 10^4 \log q)$, then we have $1/2 \le \alpha \le 1$,

$$N'(\alpha,T,q) \le \left(\frac{250359}{\log qT} + 5700\right)(q^3T^4)^{1-\alpha}(\log qT)^{6\alpha}.$$

Recently, Zhang Wenpeng]3] has improved the constants 250359 and 5700 to 138001 and 2044, respectively.

Let p_n be the n-th prime, $d_n = p_{n+1} - p_n$. By using the estimation of $N^*(\sigma,T)$ due to Heath-Brown, Cai Tianxian [3] proved that

$$\sum_{\substack{p_n \le x \\ d_n > x^\lambda}} d_n \ll x^{f(\lambda)+\varepsilon},$$

where ε is any positive number, and

$$f(\lambda) = \begin{cases} 11/10-3\lambda/5, & 1/6 < \lambda \le 7/32, \\ 1-\lambda/7, & 7/32 < \lambda \le 7/24, \\ (-68\lambda^2-28\lambda+147)/7(21-4\lambda), & 7/24 < \lambda \le 35/108, \\ 23/18-\lambda, & 35/108 < \lambda \le 31/72, \\ 11/10-3\lambda/5, & 31/72 < \lambda \le 4/9, \\ 3(1-\lambda)/2, & 4/9 < \lambda \le 5/9. \end{cases}$$

Cai's result is a refinement of Cook's.

Wang Yuan, Xie Shenggant and Yu Kunrui [1,2], and Qi Minggao [1] also obtained some results on the differences of primes. For example, Qi proved that: let $\ell \ge 1$, $R \ge 1$, $2|R$, $(\ell, R) = 1$, and p_i denote the i-th prime in the arithmetic progression $\ell+dk$, $d = 0,1,2,\ldots$. and let $r \ge 1$

$$E_r = \liminf_{j\to\infty} \frac{p_{j+r}-p_j}{\phi(R)\log p_j}.$$

Then we have

$$E_r \le \frac{2r-1}{16}\{4r + (4r-1)\frac{\theta_r}{\sin\theta_r}\},$$

where θ_r satisfies

$$\theta_r + \sin\theta_r = \pi/(4r).$$

This is an improvement of a result of Huxley's.

[3] Some fundamental properties of $\zeta(s)$ and $L(s,\chi)$. In 1936-1947, Wang Fuchun [1-6] obtained many results for $\zeta(s)$, especially he proved some important mean value theorems of $\zeta(s)$. In 1956-1958, Min Sihe [1,2,3], Min and Yin Wenlin [1] investigated a generalization of $\zeta(s)$:

$$Z_{n,k}(s) = \sum_{x_1=-\infty}^{\infty} \cdots \sum_{x_k=-\infty}^{\infty} \frac{1}{(x_1^n+\ldots+x_k^n)^s},$$

where $2|n > 0$, and obtained many interesting properties for it. Recently, Zhang Nanyue [1-5], Zhang Nanyue and Zhang Shunyan [1,2], and Zhang Shunyan [1] obtained some new results on $\zeta(s)$ and some new proofs for some fundamental properties of $\zeta(s)$. For example, Zhang Nanyue [5] obtained an integral representation of $\zeta(s)$ in the form

$$\Gamma(s)\zeta(s)\sin\frac{\pi(1-s)}{4} = 2^{(s-3)/2}\int_0^\infty \left(\frac{\text{sh}x-\sin x}{\text{ch}x-\cos x} - \theta(s)\right)x^{s-1}dx,$$

where $\theta(s) = 0$, if $-1 < \sigma < 0$; $\theta(s) = 1$, if $\sigma > 0$. and then, using this, he gave three derivations of the functional equation of $\zeta(s)$. Recently, Wang Wei [4] improved Rane's result, and proved that

$$\sum_{\chi \bmod q}^{*} \int_0^T |L(\tfrac{1}{2}+it,\chi)|^4 dt = T\sum_{\ell=0}^{4} a_\ell(\log \tfrac{qT}{2\pi})^\ell + O((qT)^\varepsilon \mathrm{Min}(q^{9/8}T^{7/8}, qT^{11/12})),$$

where "*" indicates the sum over primitive characters, ε is any given positive number, and

$$a_\ell = O(q^{1+\varepsilon}),\ \ell = 0,1,2,3,\quad a_4 = \frac{\phi(\phi(q))}{2\pi^2} \prod_{p|q} (1-\tfrac{1}{p})^4(1-\tfrac{1}{p^2})^{-1}.$$

Zhang Wenpeng [2,4] obtained some asymptotic formulas for the sum

$$\sum_{\chi \bmod q} |L(\tfrac{1}{2}+it,\chi)|^2$$

and the integral

$$\int_0^1 |\zeta_1(\tfrac{1}{2}+it,\alpha)|^2 d\alpha,$$

where $\zeta_1(s,\alpha) = \zeta(s,\alpha) - \alpha^s$, $\zeta(s,\alpha)$ is Hurwitz zeta function, $0 < \alpha \le 1$. These results are some improvements of Balasubramanian's and Rane's results. For example, he proved that: for $q \ge 3$, $t \ge 3$, then

$$\sum_{\chi \bmod q} |L(\tfrac{1}{2}+it,\chi)|^2 = \frac{\phi^2(q)}{q}\left\{\log(\tfrac{qt}{2\pi}) + 2\gamma + \sum_{p|q} \frac{\log p}{p-1}\right\}$$

$$+ O(qt^{-1/4}(\log qt)^{1/2} + q^{(1+\varepsilon)/2}t^{5/12} + t^{3/4}q^\varepsilon),$$

where ε is any given positive number.

[4] Goldbach numbers. A positive integer which is a sum of two odd primes is called a Goldbach number. In 1951, Linnik [3] first investigated the following problem by the use of the circle method. Find a function $f(x)$ such that the interval $[x, x+f(x)]$ contains at least one Goldbach number for $x \ge 2$. Using his method, Pan Chengdong [2], Wang Yuan [15] and Prachar [1] obtained further results concerning this problem. Another method of studying this problem is to use Selberg's inequality [1]. By Selberg's method, Kátai [1] proved that, under the RH, one has

(58) $$f(x) \ll \log^2 x;$$

and Montgomery and Vaughan [1] proved that, if the zero-density estimation of the ζ-function

(59) $$N(\alpha,T) \ll T^{c_1(1-\alpha)} \log^{c_2} T, \quad 1/2 \le \alpha \le 1, \quad T \ge 2,$$

holds, then

(60) $$f(x) \ll \chi^{(1-c_1^{-1})(1-2c_1^{-1})+\varepsilon},$$

ε being any small positive number. Using Selberg's method, Pan Chengdong [8] improved Montgomery and Vaughan's result (60); Pan showed that: if the estimation (59) holds, and if there are α_0 $(1/2 < \alpha_0 < 1)$, $c_3 > 0$, and c_4 such that the estimation

(61) $$N(\alpha,T) \ll T^{(2-c_3)(1-\alpha)} \log^{c_4} T, \quad \alpha_0 \le \alpha \le 1, \quad T \ge 2,$$

holds, then we have

$$f(x) \ll \chi^{(1-c_1^{-1})(1-2c_1^{-1})} \log^{c_5} x,$$

where c_5 is a constant depending on c_1, c_2 adn α_0. Recently, using Selberg's method Lu Minggao [3] improved the value of c_5, and proved that: if the estimation (59) with $c_1 = 2$ and $c_2 = 1$ holds, then we have

(63) $$f(x) \ll (\log x)^{7+\varepsilon},$$

where ε is any given positive number. This is an improvement of Wang's result [15] obtained by Linnik's method.

Finally, by using Selberg's method, Wang Yuan and Shan Zun [1] obtained a conditional result concerning Goldbach numbers in arithmetic progressions, which is a generalization of Kátai's result (58).

III. In this part we'll discuss some results concerning mean values of arithmetic functions.

[1] Let $p(n)$ be the least prime factor of n, $P(n)$ the greatest prime factor of n, $\beta(n) = \sum_{p|n} p$, $B(n) = \sum_{p^e \| n} ep$, and $B_1(n) = \sum_{p^e \| n} p^e$. Many results on the mean values of these arithmetic functions have been obtained by Erdös, Alladi, De Koninck, Van Lint and Ivić since 1977. Some of these

results were improved and generalized by Chinese mathematicians.

Xuan Tizou [1,2,3,4] proved the following results:

$$\sum_{2\le n\le x} \frac{B(n)-\beta(n)}{p^r(n)} = x \exp\left\{-(2r \log x \log_2 x)^{1/2} - \left(\frac{r \log x}{2\log_2 x}\right)^{1/2} \log_3 x + O\left(\left(\frac{\log x}{\log_2 x}\right)^{1/2}\right)\right\},$$

where $r > 0$, $\log_{k+1} x = \log(\log_k x)$. From (64) he derived asymptotic formuale for $\sum_{2\le n\le x}\left(\frac{1}{\beta^r(n)}-\frac{1}{B^r(n)}\right)$, $\sum_{2\le n\le x}\frac{B^r(n)}{\beta^r(n)}$, and $\sum_{2\le n\le x}\frac{\beta^r(n)}{B^r(n)}$; e.g.

$$\sum_{2\le n\le x}\left(\frac{1}{\beta^r(n)}-\frac{1}{B_1^r(n)}\right) = x \exp\left\{-((2r+1)\log x \log_2 x)^{1/2} - \left(\frac{2r+1}{4}\frac{\log x}{\log_2 x}\right)^{1/2} \log_3 x + O\left(\left(\frac{\log x}{\log_2 x}\right)^{1/2}\right)\right\},$$

where $r > 0$. From this he obtained the asymptotic formula for $\sum_{2\le n\le x}\frac{1}{B_1^r(n)}$;

$$x^r \ll \sum_{2\le n\le x}\frac{B_1^r(n)}{p^r(n)} \ll x^r,\quad x^r \ll \sum_{2\le n\le x}\frac{B_1^r(n)}{\beta^r(n)} \ll x^r,$$

$$\frac{x^r}{\log^r x} \ll \sum_{2\le n\le x}\frac{B_1^r(n)}{B^r(n)} \ll \frac{x^r}{\log^r x},\quad r > 1;$$

$$\sum_{2\le n\le x}\frac{B_1^r(n)}{g^r(n)} = x+O\left(\frac{x}{\log x}\right),\quad 0 < r < 1,$$

$$\sum_{2\le n\le x}\frac{g^r(n)}{B_1^r(n)} = x+O\left(\frac{x}{\log x}\right),\quad r > 0,$$

where $g(n)$ is $P(n)$, or $B(n)$, or $\beta(n)$; and

$$\sum_{2\le n\le x}\frac{\sigma(n)}{P(n)} = \frac{\pi^2}{12}x\left(1+O\left(\left(\frac{\log_2 x}{\log x}\right)^{1/2}\right)\right)\sum_{2\le n\le x}\frac{1}{P(n)},$$

$$\sum_{2\le n\le x}\frac{\phi(n)}{P(n)} = \frac{3}{\pi^2}x\left(1+O\left(\left(\frac{\log_2 x}{\log x}\right)^{1/2}\right)\right)\sum_{2\le n\le x}\frac{1}{P(n)}.$$

Jia Chaohua [2] proved that for any given positive integer K,

$$\sum_{2\le n\le x} \frac{p(n)}{P(n)} = x \sum_{k=1}^{K} \frac{a_k}{(\log x)^k} + O\left(\frac{x}{(\log x)^{k+1}}\right),$$

where a_k's are computable constants, especially $a_1 = 1$, $a_2 = 3$, and $a_3 = 15$.

Cai Tianxian [1] obtained the following asymptotic formulas:

$$\sum_{2\le n\le x} \frac{P(n)}{p(n)} = \left(\sum_{m=2}^{\infty} \frac{1}{m^2 p(m)}\right)\left(\sum_{k=1}^{K} \frac{(k-1)!}{2^k} \frac{1}{\log^k x}\right)x^2$$

$$+\left\{\sum_{k=2}^{K} \frac{1}{\log^k x} \sum_{i=1}^{k-1} (-1)^i \left[\sum_{\frac{k-1}{k-i}\le \ell\le j\le k-1} (-1)^{\ell} C_j^{\ell} C_{\ell(k-i)}^{i}\right] \times\right.$$

$$\left.\times \frac{(k-i-1)!}{2^{k-i}} \sum_{m=2}^{\infty} \frac{\log^i m}{m^2 p(m)}\right\} x^2 + O\left(\frac{x^2}{(\log x)^{k+1}}\right),$$

$$\sum_{2\le n\le x} P(n) = \sum_{k=1}^{K}\left[\sum_{j=0}^{k-1} \frac{(-\lambda)^j}{j!}\zeta^{(j)}(2)\right]\frac{(k-1)!}{2^k} \frac{x^2}{\log^k x} + O\left(\frac{x^2}{(\log x)^{k+1}}\right),$$

K being any given positive integer, and

$$\sum_{2\le n\le x} P(n) = \sum_{2\le n\le x} \beta(n) + O(x^{3/2}) = \sum_{2\le n\le x} B(n) + O(x^{3/2}) = \sum_{2\le n\le x} B_1(n) + O(x^{3/2}).$$

Recently, Cai [2] has obtained similar asymptotic formulas for

$$\sum_{x\ge n\equiv \ell (\mathrm{mod}\ q)} P(n) \quad \text{and} \quad \sum_{x\ge n\equiv \ell (\mathrm{mod}\ q)} \frac{P(n)}{p(n)}.$$

Li Hongze [1] proved that for any given integer K,

$$\sum_{n\le x} p(n) = \sum_{k=1}^{K} \frac{(k-1)!}{2^k} \frac{x^2}{\log^k x} + O\left(\frac{x^2}{(\log x)^{k+1}}\right).$$

Yu Xiuyuan [2] proved that

$$\sum_{n\le x} \sum_{p\mid n} \frac{1}{p} = x \sum_{p} \frac{1}{p^2} + O(x^{4/7} \log^2 x),$$

and

$$\sum_{n\le x}\sum_{p|n}\frac{1}{p}\Lambda\left(\frac{n}{p}\right)= x\sum_p \frac{1}{p^2}+O(xe^{-c\sqrt{\log x}}),$$

where c is a positive constant.

[2] Ton Kwangcheng did much work on the mean values of divisor function, and his work can be found in Hua's monograph [3]. Recently, Yang Zhaohua [2,4] obtained some lower bounds for integral mean values of the error terms of certain weighted sums for some classes of arithmetic functions. For example, let q_1, q_2, h_1, h_2, and n be positive integers, $h_1 \le q_1$, $h_2 \le q_2$, and

$$a(n) = \sum_{(q_1m_1+h_1)(q_2m_2+h_2)=n} 1,\quad f(s) = (q_1q_2)^{-s}\zeta(s;h_1/q_1)\zeta(s;h_2/q_2),$$

where $\zeta(s;a)$ is Hurwitz's ζ-function. And let $A_0(x) = \sum_{n\le x} a(n)$,

$$S_0(x) = \frac{x}{q_1q_2}\left\{\log\frac{x}{q_1q_2} - [\frac{\Gamma'}{\Gamma}\left(\frac{h_1}{q_1}\right) + \frac{\Gamma'}{\Gamma}\left(\frac{h_2}{q_2}\right) + 1]\right\}.$$

He [4] proved that for any $\lambda \ge 1$, $x \ge 4$, we have

$$\left\{\frac{1}{x}\int_1^x |A_0(y) - S_0(y)|^\lambda dy\right\}^{1/\lambda} \ge cx^{1/4},$$

where c is a positive constant independent of λ and x.

In addition, Yang [3] also obtained a result concerning the order-free integers (mod m).

Let $d_k(n)$ be the number of ways that n can be written as a product of k factors, and $D_k(x) = \sum_{n\le x} d_k(n)$. It is well known that there is a polynomial $P_{k-1}(y)$ of degree $k-1$ such that

$$D_k(x) - xP_{k-1}(\log x) = \Delta_k(x) \ll x^{1-1/k}\log^{k-2}x.$$

Letting

$$\beta_k = \inf\left\{\beta : \int_2^x (\Delta_k(y))^2 dy \ll x^{1+2\beta}\right\},$$

Zhang Wenpeng [1] proved that $\beta_5 \le 0.45$.

[3] A positive integer n is called k-full if $p|n$ implies $p^k|n$,

and 2-full integer is also called square-full integer. Let

$$f_k(n) = \begin{cases} 1, & n \text{ is } k\text{-full}, \\ 0, & \text{otherwise}. \end{cases}$$

Under the Riemann Hypothesis, Zhan Tao [1] proved that for any given integer m,

$$\int_1^x (\Delta_{2,m}(y))^2 y^{-2m-6/5} dy \sim c_m \log x, \quad x \to \infty,$$

where c_m is a positive constant, and

$$\Delta_{2,m}(y) = \sum_{n \le y} f_2(n) n^m - \frac{\zeta(3/2)}{\xi(3)} \frac{x^{m+1/2}}{2m+1} - \frac{\zeta(2/3)}{\zeta(2)} \frac{x^{m+1/3}}{3m+1}.$$

Assuming the Lindelöf Hypothesis, Ivic proved that there exists constants $c_{j,k}$ $(0 \le j \le k-1)$ such that

$$\Delta_k(x) = \sum_{n \le x} f_k(n) - \sum_{j=0}^{k-1} c_{j,k} \chi^{1/(k+j)} \ll \chi^{1/2k+\varepsilon},$$

ε being any positive number. Now let

$$a_k = (k-1)(3k^2-k)^{-1}, \quad 2 \le k \le 4,$$

$$a_k = r_k(r_k+1)^{-1}(2k+r_k)^{-1}, \quad k \ge 5,$$

where

$$r_k = [(1+\sqrt{8k+1})/2].$$

Under the Riemann Hypothesis, Zhan Tao [2] proved that (a) if $2 \le k \le 4$ or $k \ne (m^2-m)/2$, $m \ge 4$, we have

$$\int_1^x (\Delta_k(y))^2 y^{-2a_k-1} dy \sim d_k \log x, \quad x \to \infty,$$

where d_k is a positive constant; and (b) if $k = (m^2-m)/2$, $m \ge 4$, we have

$$\int_1^x (\Delta_k(y))^2 y^{-2a_k-1} dy \ll \log^3 x,$$

where the implied constant depends on k. Recently Zhan Tao [3] has proved that assuming the Riemann Hypothesis, we have

$$\int_1^x |\Delta_3(y)|dy \ll x^{17/16+\varepsilon},$$

$$\int_1^x |\Delta_4(y)|dy \ll x^{21/20+\varepsilon}.$$

He has also obtained similar results for $k \geq 5$.

In conclusion, some books on analytic number theory have been written by Chinese mathematicians. Besides Hua's famous monographs: Additive Theory of Prime Numbers [1]; Introduction to Number Theory [2]; Die Abschätzung von Exponentialsummen und ihre Anwendung in der Zahlentheorie [3], there are: (i) Min Sihe's book Methods in Number Theory [6] which is a graduate textbook; (ii) Pan Changdong and Pan Chengbiao's book Goldbach Conjecture [1] which provides a systematic exposition of methods and results, particularly those by Chinese mathematicians, concerning the Goldbach conjecture; (iii) Goldbach Conjecture edited by Wang Yuan [16]. The aim of Wang's book is to use a collection of original papers (showing the progress in techniques) to help the reader understand the major steps in the study of the Goldbach Conjecture; (iv) Elementary Proof of Prime Number Theorem written by Pan Chengdong and Pan Chengbiao [2]. A proof of PNT is called "elementary" if the theory of integral functions is not used. In this book which is written for undergraduate students, seven types of proofs selected from all the proofs published before 1983 are introduced.

References

R.C. Baker

[1] The greatest prime factor of the integers in an interval, Acta Arith. 47 (1986), 193-231.

M.B. Barban

[1] New applications of the "great sieve" of Ju. V. Linnik, Trudy Inst. Mat. Akad. Nauk USSR, 22 (1961), 1-20.

[2] The "density" of the zeros of Dirichlet L-series and the problem of the sum of primes and "near primes," Mat. Sb., 61 (1963), 418-425 (see Wang Yuan [16, 205-215]).

E. Bombieri

[1] On the large sieve, Mathematika, 12 (1965), 201-225, (see Wang Yuan [16, 227-252]).

V. Brun

[1] La série 1/5 + 1/7 + 1/11 + 1/13 + 1/17 + 1/19 + 1/29 + 1/31 + 1/41 + 1/43 + 1/59 + 1/61 + ... où les dénominateurs sont "nombres premiers jumeaux" est convergent ou finite, Bull. Sci. Math (2) (43 (1919), 100-104; 124-128.

[2] Le crible d'Eratosthène et le théorme de Goldbach, Skr. Norske Vid. Akad. Kristiania, I, 1920, no. 3, 1-36 (see Wang Yuan [16, 93-130]).

A.A. Buchstab

[1] New improvements in the method of the sieve of Eratosthenes, Mat. Sb. 46 (1938), 375-387 (see Wang Yuan [16, 131-147]).

[2] Sur la décomposition des nombres pairs en somme de deux composantes dont châcune est formée d'un nombre borné de facteurs premiers, Dokl. Akad. Nauk SSSR, 29 (1940), 544-548.

[3] New results in the investigation of the Goldbach-Euler problem and the problem of prime pairs, Dokl. Akad. Nauk SSSR, 162 (1965), 735-738 (see Wang Yuan [16, 216-222]).

D.A. Burgess

[1] On character sums and primitive roots, Proc. London Math. Soc., 12 (1962), 179-192.

Cai Tianxian

[1] An average estimation of a class of arithmetic functions, Kexue Tongbao, 29 (1984), 1481-1484.

[2] ____________, II, to appear.

[3] On the upper bound for the sum of differences between consecutive primes, to appear.

Chen Jingrun

[1] On the least prime in an arithmetical progression, Sci. Sin., 14 (1964), 1868-1871.

[2] On the representation of a large even integer as the sum of a prime and the product of at most two primes, Kexue Tongbao, 17 (1966), 385-386.

[3] ____________, Sci. Sin., 16 (1973), 157-176 (see Wang Yuan [16, 253-272]).

[4] On the distribution of almost primes in an interval, Sci. Sin., 18 (1975), 611-627.

[5] On the least prime in an arithmetical progression and two theorems concerning the zeros of Dirichlet's L-functions, Sci. Sin., 20 (1977), 529-562.

[6] On the representation of a large even integer as the sum of a prime and the product of at most two primes, II, Sci. Sin., 21 (1978), 421-430.

[7] On the Goldbach's problem and the sieve methods, Sci. Sin., 21 (1978), 701-739.

[8] On the distribution of almost primes in an interval, Sci. Sin., 22 (1979), 253-275.

[9] On the least prime in an arithmetical progression and two theorems concerning the zeros of Dirichlet's L-functions, Sci. Sin., 22 (1979), 859-889.

[10] On some problems in prime number theory, Séminaire de Théorie des Nombres, Paris 1979-1980, 167-170.

[11] On zeros of Dirichlet's L-functions, Sci. Sin. ser A, 29 (1986), 897-913.

B. Conrey

[1] Zeros of derivatives of Riemann's ζ-function on the critical line, J. Number Theory, 16 (1984), 49-74.

B. Conrey, A. Ghosh

[1] A simpler proof of Levinson's theorem, Math. Proc. Camb. Phil. Soc., 97 (1985), 385-395.

T. Estermann

[1] Eine neue Darstellung and neue Anwendungen der Viggo Brunschen Methode, J. Reine Angew. Math., 168 (1932), 106-116.

A. Fujii

[1] Some remarks on Goldbach's problem, Acta Arith., 32 (1977), 27-35.

S. Graham

[1] On Linnik's constant, Acta Arith., 39 (1981), 163-179.

[2] The greatest prime factor of the integers in an interval, J. London Math. Soc., 24 (1981), 427-440.

H. Halberstam

[2] A proof of Chen's theorem, Asterisque, 24-25 (1975), 281-293.

H. Halberstam, D.R. Heath-Brown, H.-E. Richert

[1] Almost-primes in short intervals, Recent Progress in Analytic Number Theory I, edited by Halberstam and Hooley, Acad. Press, 1981, 69-101.

H. Halberstam, H.-E. Richert

[1] Sieve Methods, Acad. Press, 1974.

Hua Lookeng

[1] Additive Theory of Prime Numbers, Trud. Inst. Mat. Steklov, 22 (1947); Science Press, Beijing, 1952; AMS, 1965.

[2] Introduction to Number Theory, Science Press, Beijing, 1957, Springer-Verlag, 1982.

[3] Die Abschätzung von Exponentialsummen and ihre Anwendung in der Zahlentheorie, Enz. der Math. Wiss, I, 2, Heft 13, Teil 1, Leipzig, Teubner, 1959,; Science Press, Beijing, 1963.

H. Iwaniec

[1] Rosser's sieve - bilinear forms of the remainder terms - some applications, Recent Progress in Analytic Number Theory I, Acad. Press, 1981, 203-230.

H, Iwaniec, M. Laborde

[1] P_2 in short intervals, Ann. Inst. Fourier, Grenoble 31 (1981), 37-56.

Jia Chaohua

[1] The greatest prime factor of the integers in a short interval I, Acta Math. Sin., 29 (1986), 815-825.

[2] A generalization of prime number theoreum, Chinese Adv. Math., to appear.

[3] The greatest prime factor of the integers in a short interval II, to appear.

W.B. Jurkat, H.-E. Richert

[1] An improvement of Selberg's sieve method I, Acta Arith., 11 (1965), 217-240.

J. Jutila

[1] A new estimate for Linnik's constant, Ann. Acad. Sci. Fennicae, 471 (1970), 8 pages.

[2] On the Linnik's constant, Math. Scand. 41 (1977), 54-62.

I. Kátai

[1] A comment on a paper of Ju. V. Linnik, Magyer Tud. Akad. Mat. Fiz. Oszt. Közl, 17 (1967), 99-100.

P. Kuhn

[1] Neue Abschätzungen auf Grund der Viggo Brunschen Siebmethode, 12 Skand. Mat. Kongr., Lund, 1953, 160-168.

[2] Über die Primteiler eines Polynoms, Proc. Inter. Congr. Math., Amsterdam, 1954, 35-37 (see Wang Yuan [16, 148-150]).

N. Levinson

[1] More than one third of the zeros of Riemann's zeta-function are on $\sigma = 1/2$, Adv. Math., 13 (1974), 383-436.

[2] A simplification of the proof that $N_0(T) > (1/3)N(T)$ for Riemann's zeta-functon, Adv. Math., 18 (1975), 239-242.

[3] Deduction of semi-optimal mollifier for obtaining lower bounds for $N_0(T)$ for Riemann's zeta-function, Proc. Nat. Acad. Sci. USA, 72 (1975), 294-297.

Li Hongze

[1] The mean value of the arithmetic function $p(n)$, to appear.

Ju. V. Linnik

[1] The large sieve, Dokl. Akad. Nauk SSSR, 30 (1941), 290-292.

[2] On the least prime in an arithmetic progression I: The basic theorem, Mat. Sb., 15 (1944), 139-178; II: The Deuring-Heilbronn's phenomenon, Mat. Sb., 15 (1944), 347-368.

[3] Some conditional theorems concerning binary Goldbach problem, Izv. Akad. Nauk SSSR, Ser. Mat., 16 (1952), 503-530.

[4] An asymptotic formula in an additive problem of Hardy-Littlewood, ibid., 24 (1960), 629-706.

Liu Hongquan

[1] The asymptotic formula for $F_{2^a}(x)$, Acta Math. Sin., to appear.

Lou Shitou

[1] A lower bound for the number of zeros of Riemann's zeta-function on $\sigma = 1/2$, Recent Progress in Analytic Number Theory I, edited by Halberstam and Hooley, Acad. Press, 1981, 319-324.

Lou Shitou, Yao Qi

[1] Lower bound for zeros of Riemann's zeta-function on $\sigma = 1/2$, Acta Math. Sin., 24 (1981), 390-400.

Lu Minggao

[1] An inequality involving trigonometrical polynomials, Kexue Tongbao, 27 (1982), 1151-1156.

[2] A conditional result on the least positive primitive root, Kexue Tongbao.

[3] On the Goldbach number, Sci. Sin. 27 (1984), 242-252.

[4] The maximum number of mutually orthogonal Latin squares, Kexue Tongbao, 30 (1985), 154-159.

[5] The asymptotic formula for $F_2(x)$, Sci. Sin., to appear.

Min Sihe (Min Szu-Hao)

[1] On a way of generalization of the Riemann ζ-function I, Acta Math. Sin., 5 (1955), 285-294.

[2] A generalization of Riemann's ζ-function II, ibid, 6 (1956), 1-12.

[3] On a generalization of Riemann's ζ-function III, ibid., 6 (1956), 347-362.

[4] On the non-trivial zeros of Riemann's ζ-function, Acta Sci. Nat. Univ. Pekinensis, 2 (1956), 165-190.

[5] Remarks on $\pi(x)$ and $\zeta(s)$, ibid., 2 (1956), 297-302.

[6] Methods in Number Theory, I, II, Science Press, Beijing, 1981.

Min Sihe, Yin Wenlin

[1] On the mean-value theorems of $Z_{n,k}(s)$, Acta Sci. Nat. Univ. Pekinensis, 4 (1958), 50-64.

Mo Guoduan

[1] Evaluation of a class of integral in the theory of Riemann's zeta-function, Acta Math. Sin., 28 (1985), 684-696.

H.L. Montgomery, R.C. Vaughan

[1] The exceptional set in Goldbach's problem, Acta Arith., 27 (1975), 353-370.

Eugene K.-S. Ng

[1] On the number of solutions of $N - p = p_3$, J. Number Theory, 18 (1984), 229-237.

[2] On the sequences $N - p$, $p + 2$ and the parity problem, Arch. Math. 42 (1984), 430-438.

Pan Chengbiao

[1] A simplification of the proof of Levinson's theorem, 22 (1979), 343-353.

[2] Number theory in China, La Teoria dei Numeri Nella Cina Antica e Di Oggi, Ferrara, Maggio, 1979, 1-40.

[3] The weighted sieve method and the mean value theorem, ibid., 57-79.

[4] On the upper bound of the number of ways to represent an even integer as a sum of two primes, Sci. Sin., 23 (1980), 1368-1377.

Pan Chengdong

[1] On the least prime in an arithmetical progression, Acta Sci. Nat. Univ. Pekinensis, 1957, no. 1, 1-34; Sci. Record (N.S.), 1 (1958), 311-313.

[2] Some new results on additive theory of prime numbers, Acta Math. Sin., 9 (1959), 315-329.

[3] On representation of even numbers as the sum of a prime and an almost prime, Acta Math. Sin., 12 (1962), 95-106; Sci. Sin., 11 (1962), 873-888 (see Wang Yuan [16, 192-204]).

[4] On representation of large even integer as the sum of a prime and a product of at most four primes, Acta Sci. Nat. Univ. Shandong, 1962, no. 2, 40-62; Sci. Sin., 12 (1963), 455-473.

[5] A note on the large sieve method and its applications, Acta Math. Sin. 13 (1963), 262-268.

[6] A new application of Linnik's large sieve, Acta Math. Sin., 14 (1964), 597-608; Sci. Sin., 13 (1964), 1045-1053.

[7] On the zeros of the zeta-function of Riemann, Sci. Sin., 14 (1965), 303-305.

[8] On Goldbach number, Kexue Tongbao, Special Ser. Math. Phy. Chem., 1980.

[9] A new mean value theorem and its applications, <u>Recent Progress in Analytic Number Theory I</u>, edited by Halberstam and Hooley, Acad. Press, 1981, 275-288 (see Wang Yuan [16, 273-285]).

Pan Chengdong, Ding Xiaqi

[1] A mean value theorem, Acta Math. Sin., 18 (1975), 254-262; 19 (1976), 217-218.

[2] A new mean value theorem, Sci. Sin. Special Issue (II), 1979, 149-161.

Pan Chengdong, Ding Xiaqi, Wang Yuan

[1] On the representation of every large even integer as a sum of a prime and an almost prime, Sci. Sin., 18 (1975), 599-610.

Pan Chengdong, Pan Chengbiao

[1] Goldbach Conjecture, Science Press, Beijing, 1981.

[2] Elementary Proofs of Prime Number Theorem, Shanghai Sci. Tech. Press, Shanghai, 1987.

K. Prachar

[1] Über die Anwendung einer Methode von Linnik, Acta Arith., 29 (1976), 367-376.

Qi Minggao

[1] On the differences of primes in arithmetic progressions, J. Qinghua Univ., 21 (1981), 25-36.

D.A. Rawsthorne

[1] The linear sieve, revisited, Acta Arith., 44 (1984), 181-190.

A. Rényi

[1] On the representation of even number as the sum of a prime and an almost prime, Izv. Akad. Nauk SSSR, Ser. Mat. 12 (1948), 57-78 (see Wang Yuan [16, 163-169]).

H.-E. Richert

[1] Selberg sieve with weights, Mathematika, 16 (1969), 1607-1624.

H. Riesel, R.C. Vaughan

[1] On sums of primes, Arkiv für Math. 21 (1983), 45-74.

K.A. Rodosskii

[1] On the least prime number in an arithmetic progression, Mat. Sb., 34 (1954), 331-356.

P.M. Ross

[1] On Chen's theorem that each large even number has the form $p_1 + p_2$ or $p_1 + p_2p_3$, J. London Math. Soc. (2), 10 (1975), 500-506.

K.F. Roth

[1] On the large sieve of Linnik and Rényi, Mathematika, 12 (1965), 1-9.

A. Selberg

[1] On the normal density of primes in small intervals, and the difference between consecutive primes, Arch. Math. Naturvid 47 (1943), no. 6, 87-105.

[2] On an elementary method in the theory of prime, Norske Vid. Selsk. Forh. Trondhjem, 19 (1947), 64-47 (see Wang Yuan, [16, 151-154]).

[3] On elementary methods in prime number theory and their limitations, 11 Skand. Mat. Kongr. Trondhjem, 1949, 13-22.

Shao Pintsung

[1] On the distribution of the value of a class of arithmetical functions, Acta Sci. Nat. Univ. Pekinensis, 1956, no. 3, 261-278.

[2] ____________, Bull. Acad. Polon. Sci. CILIII, 4 (1956), 569-572.

[3] On a problem of Schinzel, Chinese Adv. Math., 2 (1956), 703-710.

[4] A note on some properties of arithmetical functions $\omega(n)$ and $\Omega(n)$, Acta Math. Sin., 23 (1980), 758-762.

[5] On the divisor problem of Erdös, Acta Math. Sin., 24 (1981), 797-800.

Shao Xiong

[1] On the lower bound of the number of solutions of $N - p = P_3$ (II), Acta Math. Sin. 30 (1987), 125-131.

Shen Zun

[1] On the Diophantine equation $\sum_{i=0}^{k} \frac{1}{x_i} = \frac{a}{n}$, China Ann Math., B. 7 (1986), 213-220.

R.C. Vaughan

[1] On the estimation of Schnirelmann's constant, J. Reine Angew. Math., 290 (1977), 93-108.

A.I. Vinogradov

[1] The density hypothesis for Dirichlet L-series, Izv. Akad. Nauk SSSR, Ser Mat., 29 (1965), 903-934; Corrigendum, ibid., 30 (1966), 719-720 (see Wang Yuan, [16, 223-226]).

Wang Fuchun (Wang Fu Traing)

[1] A remark on the mean value theorem of Riemann's zeta function, Sci. Report Tôhoku Imperial Univ., (1) 25 (1936), 381-391.

[2] On the mean value theorem of Riemann's zeta-function, ibid., (1) 2 (1936), 392-414.

[3] A note on zeros of Riemann zeta-function, Proc. Imp. Acad. Tokyo, 12 (1937), 305-306.

[4] A formula on Riemann zeta-function, Ann. Math., (2) 46 (1945), 88-92.

[5] A note on the Riemann zeta-function, Bull. Amer. Math. Soc. 52 (1946), 319-321.

[6] A mean value theorem of the Riemann zeta-function, Quart. J. Math., Oxford ser., 18 (1947), 1-3.

Wang Wei

[1] On two theorems of Linnik, Kexue Tongbao, 1984, no. 2, 765.

[2] On the distribution of zeros of Dirichlet L-functions, Acta Sci. Nat. Univ. Shandong, 21 (1986), no. 3, 1-13.

[3] On the least prime in an arithmetic progression, Acta Math. Sin., 29 (1986), 826-836.

[4] The fourth power mean of Dirichlet's L-functions, to appear.

Wang Yuan

[1] On the representation of large even integer as a sum of a product of at most three primes and a product of at most four primes, Acta Math. Sin., 6 (1956), 500-513.

[2] On the representation of large even integer as a sum of a prime and a product of at most four primes, Acta Math. Sin., 6 (1956), 565-582.

[3] On sieve methods and some of the related problems, Sci. Record (N.S.), 1 (1957), no. 1, 9-12.

[4] On sieve methods and some of their applications, Sci. Record (N.S.), 1 (1957), no. 3, 1-5.

[5] On the representation of large even number as a sum of two almost primes, Sci. Record (N.S.), 1 (1957), no. 5, 15-19 (see [16, 155-159]).

[6] On some properties of integral valued polynomials, Chinese Adv. Math., 3 (1957), 416-423.

[7] A note on some properties of the arithmetical functions $\phi(n)$, $\sigma(n)$ and $d(n)$, Acta Math. Sin., 8 (1958), 1-11.

[8] On sieve methods and some of their applications I, Acta Math. Sin., 8 (1958), 413-429; Sci. Sin., 8 (1959), 357-381.

[9] On sieve methods and some of their applications II, Acta Math. Sin., 9 (1959), 87-100; Sci. Sin., 11 (1962), 1607-1624.

[10] On the least primitive root of a prime, Sci. Record. (N.S.), 3 (1959), no. 5, 174-179; Acta Math. Sin., 9 (1959), 432-441; Sci. Sin., 10 (1961), 1-14.

[11] On the representation of large integer as a sum of a prime and an almost prime, Acta Math. Sin., 10 (1960), 168-181; Sci. Sin., 11 (1962), 1033-1054 (see [16, 170-191]).

[12] A note on the maximal number of pairwise orthogonal Latin square of a given order, Sci. Sin., 13 (1964), 841-843.

[13] On the maximal number of pairwise orthogonal Latin square of order s, Acta Math. Sin., 16 (1966), 400-410.

[14] A note on the theorem of Davenport, Acta Math. Sin., 18 (1975), 286-289.

[15] On Linnik's method concerning the Goldbach number, Sci. Sin., 20 (1977), 16-30.

[16] Goldbach Conjecture, edited by Wang Yuan, World Scientific Publ. Co., Singapore, 1984.

Wang Yuan, Shan Zun

[1] A conditional result on Goldbach problem, Acta Math. Sin. New Ser., 1 (1985), 72-78.

Wang Yuan, Xie Shenggang (Hsieh Shengkang), Yu Kunrui

[1] Two results on the distribution of prime numbers, J. China Univ. Sci. Tech., 1 (1965), 32-38.

[2] Remarks on the difference of consecutive primes, Sci. Sin., 14 (1965), 786-788.

Xie Shenggang

[1] On the distribution of 3-twin primes, Chinese Adv. Math., 8 (1965).

[2] On the k-twin primes problem, Acta Math. Sin., 26 (1983), 378-384.

[3] The general twin primes problem, Chinese Adv. Math., 12 (1983), 313-320.

[4] The linear combinatorial sieve, ibid., 13 (1984), 119-144.

[5] The estimation of an important constant in sieve method, J. China Univ. Sci. Tech. Math. Issue, 1985, 100-105.

[6] On equation $ap - bP_2 = m$, Acta math. Sin. New Ser. 3 (1987), 54-57.

Xuan Tizou

[1] Sums of certain large additive functions, J. Beijing Normal Univ. Nat. Sci., 1984, no. 2, 11-18.

[2] Sums of reciprocals of a class of additive functions, J. Math. (PRC), 5 (1985), 33-40.

[3] On the asymptotic formulae for power of quotient of certain arithmetical functions, J. Beijing Normal Univ. Nat. Sci., 1986, no. 1, 1-10.

[4] Estimates of certain sums involving the largest prime factor of an integer, ibid., (to appear).

Yang Zhaohua

[1] An improvement for a theorem of Davenport's, Kexue Tongbao, 26 (1984), no. 10.63; J. China Univ. Sci. Tech., 15 (1985), 1-5.

[2] Integral average order estimation of error term of weighted sum for a class of arithmetical functions, J. China Univ. Sci. Tech. Math. Issue, 1985, 106-117.

[3] A note for order-free integer (mod m), J. China Univ. Sci. Tech., 16 (1986), 116-118.

[4] A divisor problem in arithmetic progression, Acta Math. Sin., to appear.

Yin Wenlin

[1] Remarks on the representation of large integers as sum of primes, Acta Sci. Nat. Univ. Pekinensis, 1956, no. 3, 323-326.

[2] Note of the representation of large integers as sum of primes, Bull. Acad. Polon. Sci. CI III, 4 (1956), 793-795.

[3] On Schnirelman density, Acta Sci. Nat. Univ. Pekinensis, 1956, no. 4, 401-410.

[4] An application of the mean value theorem of the Dirichlet series, Acta Sci. Nat. Univ. Pekinensis, 1957, no. 4, 391-394.

Yu Xiuyuan

[1] On some properties of L-functions character modulus p^n, Chinese Ann. Math., 2 (1981), 377-386.

[2] An estimate on the distribution of weakly compositive numbers, Acta Sci. Nat. Univ. Shandong, (to appear).

Zhan Tao

[1] On the error function of the square-full integers, Chinese Adv. Math., 15 (1986), 220-221.

[2] The distribution of K-full integers, to appear.

[3] ____________, II, to appear.

[4] Bombieri's theorem in short intervals, to appear.

Zhang Dexian

[1] A mean value theorem for function $\mu(n)$, J. Shandong Col. Oce., 12 (1982), 11-20.

[2] Formula of the large sieve with single prime power, ibid, 14 (1984), 91-98.

[3] The extension of Barban's theorem, ibid., to appear.

Zhang Mingyao

[1] On the estimate of Schnirelman's constant, Kexue Tongbao, 29 (1984), 565; Acta Sci. Nat. Univ. Anhui, 1984, No. 2, 14-21.

[2] On the finite groups of a given order, Kexue Tongbao, to appear.

[3] Goldbach conjecture and a parity problem, Research Memorandum, Inst. math. Acad. Sin., No. 27, 1986; Kexue Tongbao, to appear.

[4] Some applications of Brun's and Selberg's sieve methods, to appear.

[5] On a generalization of Goldbach conjecture, to appear.

Zhang Mingyao, Ding Ping

[1] An improvement of the estimate of Schnirelman's constant, Kexue Tongbao, 28 (1983), 1012-1017; J. China Univ. Sci. Tech. Math. Issue, 1983, 31-53.

Zhang Nanyue

[1] On the Stieltjes constants of the zeta function, Acta Sci. Nat. Univ. Pekinensis, 1981, no. 4, 20-24.

[2] On the functional equation of the zeta function, ibid., 1982, no. 2, 30-33.

[3] A representation of Riemann zeta function, J. Math. Res. Exp., 2 (1982), 119-120.

[4] Ramanujan's formula and the value of Riemann zeta function at all positive odd integers, Chinese Adv. Math., 12 (1983), 61-71.

[5] The series $\sum_{n=1}^{\infty} n^{-2}e^{-z^2/n^2}$ and Riemann zeta function, Acta Math. Sin., 26 (1983), 736-744.

Zhang Nanyue, Zhang Shunyan

[1] Two consequences of Riemann hypothesis, Acta Sci. Nat. Univ. Pekinensis, 1982, no. 4, 1-6.

[2] The Weil formula of Riemann zeta function, ibid., 1984, no. 2, 12-19.

Zhang Shunyan

[1] An application of the functional equation of $\zeta(s)$, ibid., 1981, no. 2, 42-46.

Zhang Wenpeng

[1] On the divisor problem, to appear.

[2] On the mean square value of Dirichlet's L-functions, to appear.

[3] On the zero density of Dirichlet's L-functions, to appear.

[4] On the Hurwitz zeta function, to appear.

Zhang Yitang

[1] Two theorems on the zero density of the Riemann zeta function, Acta Math. Sin. New Ser., 1 (1985), 274-285.

Shandong University

Peking University

University of Science and Technology of China

Contemporary Mathematics
Volume 77, 1988

NUMBER THEORETIC METHOD IN NUMERICAL ANALYSIS*

Wang Yuan

§1. Introduction.

Due to the rapid developments and applications of computer technology, mathematicians began to take a serious interest in the applications of number theory to numerical analysis about thirty years ago. The progress achieved thus far has been both important practically and satisfactorily from the theoretical point of view. Many fields in number theory, such as the theory of diophantine approximation, uniform distribution, and algebraic number theory, have been used with some success in solving computational problems.

We shall illustrate several such applications in numerical integration and its related subjects. From the seventeenth century until now, a great deal of effort was made in developing methods for approximating single integrals and there were only a few works on multiple quadrature until the 1950s. But in the past thirty years, a number of new methods have been devised of which the number theoretic method has also proved to be very effective in studying numerical quadrature of multiple integrals.

The simplest case is to evaluate the definite integral

$$I(f) = \int_0^1 \cdots \int_0^1 f(x_1,\dots,x_s)dx_1\dots dx_s = \int_{G_s} f(\underline{\underline{x}})d\underline{\underline{x}}, \qquad (1)$$

where G_s denotes the s-dimensional unit cube and $\underline{\underline{x}} = (x_1,\dots,x_s)$. Suppose that

$$\frac{\partial^s f}{\partial x_1 \dots \partial x_s}$$

and its lower derivatives are periodic functions of s variables each with period 1 and are bounded by C.

---

*This is an expanded version of an article entitled "Applications of Number Theory to Numerical Analysis," appeared in Recent progress in Analytic Number Theory (edited by Halberstam and Hodey), Academic Press, 1981.

The most obvious approach to $I(f)$ is to regard the integral as an s-fold iterated integral and apply a one-dimensional quadrature formula to each variable respectively. For example, from the trapezoid rule for a single integral, we obtain the Cartesian product formula of trapezoid formulas. That is, we use

$$\frac{1}{n}\sum_{\ell_1=0}^{m-1}\cdots\sum_{\ell_s=0}^{m-1} f\left(\frac{\ell_1}{m},\ldots,\frac{\ell_s}{m}\right) \tag{2}$$

to approximate (1), where $n = m^s$. One can show that the error between (1) and (2) cannot be better than

$$O(Cn^{-1/s}) \tag{3}$$

which increases rapidly as s increases.

A quite new approach to multiple quadrature was developed in the 1940s as a part of the Monte Carlo method by S. Ulam and J. von Neumann. The basic idea of the Monte Carlo method is to replace an analytic problem by a probabilistic problem with the same solution, and then investigate the latter problem by statistical experiments. It has become a conventional numerical method in evaluating multidimensional integrals and standarized the integration of complicated functions that would otherwise be impossible.

Suppose that $\underline{\underline{x}} = (x^1,\ldots,x^s)$ is a random variable that is uniformly distributed on G_s. Then $f(\underline{\underline{x}})$ is a random variable whose mean value and standard deviation are given by

$$m(f(\underline{\underline{x}})) = \int_{G_s} f(\underline{\underline{x}})d\underline{\underline{x}} \quad \text{and} \quad \sigma(f(\underline{\underline{x}})) = \left((m(f(\underline{\underline{x}})^2) - m(f(\underline{\underline{x}}))^2)\right)^{1/2}.$$

If we take N independent samples $\underline{\underline{x}}_1,\ldots,\underline{\underline{x}}_N$ of $\underline{\underline{x}}$ and form the average

$$J(f,N) = \frac{1}{N}\sum_{i=1}^{N} f(\underline{\underline{x}}_i), \tag{4}$$

we might reasonably expect this average to be close to the mean of $f(\underline{\underline{x}})$. It is known that J is approximately normally distributed at least when N is large. So we may say that, for example, the probability of

$$|J(f,N) - I(f)| \le 2\sigma(f)N^{-1/2} \tag{5}$$

is about 0.95. Although the error term $O(N^{-1/2})$ in (5) is much better than that in the Cartesian product formula of trapezoid formulas, it is in the sense of probability and not the absolute error in the usual meaning.

The number theoretic method of numerical quadrature of multiple integrals is based on the theory of uniform distribution (Weyl [1]). We use $\underline{\underline{\gamma}}$ to denote a vector with real components and $|\underline{\underline{\gamma}}| = \gamma_1 \dots \gamma_s$. Let $n_1 < n_2 < \dots$ be a sequence of positive integers. Let

$$P_{n_\ell}(k) = (x_1^{(n_\ell)}(k), \dots, x_s^{(n_\ell)}(k))), \quad 1 \le k \le n_\ell$$

be a set of points in G_s. For any $\underline{\underline{\gamma}} \in G_s$, let $N_{n_\ell}(\underline{\underline{\gamma}})$ denote the number of points of $P_{n_\ell}(k)$ $(1 \le k \le n_\ell)$ satisfying the inequalities

$$0 \le x_i^{(n_\ell)}(k) < \gamma_i, \quad 1 \le i \le s.$$

Further, let

$$\sup_{\underline{\underline{\gamma}} \in G_s} \left| \frac{N_{n_\ell}(\underline{\underline{\gamma}})}{n_\ell} - |\underline{\underline{\gamma}}| \right| = D(n_\ell).$$

If $D(n) = o(1)$ (as $n \to \infty$), then the sequence of sets $(P_{n_\ell}(k))$ $(n_1 < n_2 < \dots)$ is said to be uniformly distributed with discrepancy $D(n)$, or simply that the set $P_n(k)$ $(1 \le k \le n)$ has discrepancy $D(n)$; hereafter we often omit the index ℓ.

It was shown by K.F. Roth [1] that

$$D(n) \ge c(s)\frac{(\log n)^{(s-1)/2}}{n}.$$

So if the discrepancy of $P_n(k)$ $(1 \le k \le n)$ satisfies

$$D(n) = O(n^{-1+\varepsilon}),$$

where ε is any preassigned positive number, then the set $P_n(k)$ $(1 \le k \le n)$ is said to be best uniformly distributed.

For any given set $P_n(k)$ $(1 \le k \le n)$, we may use

$$J_1(f, P_n) = \frac{1}{n} \sum_{k=1}^{n} f(P_n(k)) \tag{6}$$

to approximate the multiple integral $I(f)$. The difference between $I(f)$ and $J_1(f, P_n)$ can be shown to be dominated by

$$|I(f) - J_1(f, P_n)| \le C2^s D(n) \tag{7}$$

(Sobol [1], Hlawka [1]). Hence the problem of finding the best quadrature formula is equivalent to the problem of finding the best uniformly distributed sequence of sets.

It is surprising that the number theoretic method can be used to construct various uniformly distributed sets with discrepancy $D(n) = O(n^{-1+\varepsilon})$, which are of a best possible kind and much better than the probabilistic errors in Monte Carlo method. From the viewpoint of numerical analysis, we demand not only that the discrepancy of the set $P_n(k)$ $(1 \le k \le n)$ is small but also that the $P_n(k)$ is convenient for computation. Now we introduce the sets proposed by Korobov (1959) [1,3] and Hlawka (1962) [2] independently as follows. Let

$$Q_n(k) = \left[\left\{\frac{a_1 k}{n}\right\}, \dots, \left\{\frac{a_s k}{n}\right\}\right], \quad 1 \le k \le n, \tag{8}$$

where $\{x\}$ denotes the fractional part of x. Suppose that $n = p$ is a prime number. Then they proved that there exists an integral vector $\underline{\underline{a}} = \underline{\underline{a}}(p)$ such that the set (8) has discrepancy

$$D(n) = O(n^{-1}(\log n)^s). \tag{9}$$

Hence the set (8) is best uniformly distributed for $n = p$. Moreover, the structure of $Q_n(k)$ is quite simple and convenient to use. If the set (8) has discrepancy (9), then $\underline{\underline{a}}(n)$ is called a set of optimal coefficients modulo n or a good lattice point modulo n. In 1978, Niederreiter [1] generalized the result of Korobov and Hlawka to any given natural number n.

For any given $n = p$, Korobov [2,3] proved that it requires

$$O(n^2) \tag{10}$$

elementary operations for obtaining the vector $\underline{\underline{a}} = \underline{\underline{a}}(n)$. For large n, this can be improved to $O(n^{3/4})$. However, from the viewpoint of numerical analysis, Korobov's result is only an existence theorem. It seems that one of the central problems in numerical integration is to find a direct method for obtaining the optimal coefficients.

For $s = 2$, Hua Lookeng and Wang Yuan [1] and Bahvalov [1] proved independently in 1959 that if we take $n = F_m$, then $a_1 = 1$ and $a_2 = F_{m-1}$ is a set of optimal coefficients modulo F_m, where (F_m) is the usual Fibonacci sequence, that is, the sequence of integers defined by the recurrent formula

$$F_0 = 0, \ F_1 = 1, \ F_{m+1} = F_m + F_{m-1} \quad (m \ge 1)$$

or

$$F_m = \frac{1}{\sqrt{5}}\left[\left(\frac{1+\sqrt{5}}{2}\right)^m - \left(\frac{1-\sqrt{5}}{2}\right)^m\right], \quad m = 0,1,2,\ldots .$$

Evidently, it requires only

$$O(\log F_m) \tag{11}$$

elementary operations to obtain F_m. And we have the quadrature formula

$$\int_0^1\int_0^1 f(x_1,x_2)dx_1dx_2 - \frac{1}{F_m}\sum_{k=1}^{F_m} f\left(\frac{k}{F_m}, \left\{\frac{F_{m-1}k}{F_m}\right\}\right) = O\left(\frac{\log F_m}{F_m}\right).$$

To extend the method to the s-dimensional case $(s > 2)$ we need only to extend the number F_m. First, note that the fraction $\frac{F_{m-1}}{F_m}$ is convergent to the golden number $\frac{\sqrt{5}-1}{2} = 2\cos\frac{2\pi}{5}$. Let p be a prime ≥ 5. Consider the cyclotomic field $Q(\cos\frac{2\pi}{p})$. it has an integral basis

$$\omega_\ell = 2\cos\frac{2\pi\ell}{p}, \quad 1 \leq \ell \leq s, \quad s = \frac{p-1}{2}.$$

We can propose an algorithm for finding a sequence of sets of integers $(h_1,\ldots,h_{s-1};n)$ such that

$$|\omega_\ell - \frac{h_\ell}{n}| = O(n^{-1-(1/(s-1))}), \quad 1 \leq \ell \leq s-1. \tag{12}$$

$(1,h_1,\ldots,h_{s-1})$ is regarded to be an "optimal coefficients" modulo n, and we can show that the set

$$\left(\frac{k}{n}, \left\{\frac{h_1k}{n}\right\},\ldots,\left\{\frac{h_{s-1}k}{n}\right\}\right), \quad 1 \leq k \leq n \tag{13}$$

has discrepancy

$$O(n^{-1/2-((1/2(s-1))+\varepsilon)}), \tag{14}$$

where ε is any preassigned positive number. To obtain such a set it needs only

$$O(\log n) \tag{15}$$

elementary operations (see Hua and Wang [2,3,4,5,6], Wang, Xu and Zhang [1]).

As for the comparison of cyclotomic field method with Korobov's method for obtaining optimal coefficients, Haber [1] stated

"The Korobov's method takes An^2s seconds (to get $\underline{\underline{a}}(n)$ for all $s' \leq s$), where now A is 10^{-5} for calculations reported here (on a UNIVAC 1108). This is much better, and the method can be carried out at reasonable expense for n up to 10^4 or so and s up to 10 or 20. However, there is a reason to suspect that practical formulas for s as high as 10 will require n's of the order 10^5 or more, and again the calculation becomes excessively long."

"The third method (cyclotomic field method) requires only As^3 seconds, the only calculation of the significant length that is necessary is the solution of the linear system (12), A is apt to be about 10^{-3} or lower. The length of the calculation is thus no obstacle for s up to 100, at least, and n arbitrarily large."

§2. Diophantine approximation and uniform distribution.

Let $\mathcal{F}_s$ be an algebraic number field of degree s over the rational number field Q. Put $s = r_1 + 2r_2$. For a number α of $\mathcal{F}_s$, let $\alpha^{(1)},\dots,\alpha^{(s)}$ be its conjugates, where $\alpha^{(1)},\dots,\alpha^{(r_1)}$ are real numbers, $\alpha^{(r_1+1)},\dots,\alpha^{(r_1+r_2)}$ are complex numbers and $\alpha^{(r_1+r_2+1)} = \overline{\alpha^{(r_1+1)}},\dots,$ $\alpha^{(r_1+2r_2)} = \overline{\alpha^{(r_1+r_2)}}$.

Lemma 1. Put $r = r_1 + r_2$. Let $\gamma_1,\dots,\gamma_r$ be a given set of real numbers satisfying

$$\sum_{j=1}^{r_1}\gamma_j + 2\sum_{j=r_1+1}^{r}\gamma_j = 0. \tag{16}$$

Then there exists a unit $\eta \in \mathcal{F}_s$ such that

$$c(\mathcal{F}_s)^{-1}e^{\gamma_i} \le |\eta^{(i)}| \le c(\mathcal{F}_s)e^{\gamma_i}, \quad 1 \le i \le r,$$

where we use $c(f,\dots,g)$ to denote a positive constant depending on $f,\dots,g$ only, but not always with the same value, see e.g., Hua and Wang [6].

For a real algebraic number field $\mathcal{F}_s$, set $\gamma_1 = \gamma$ an $\gamma_2 = \dots = \gamma_r = \bar{\gamma}$. Then (16) becomes

$$\gamma + (s-2)\bar{\gamma} = 0 \quad \text{or} \quad \bar{\gamma} = \frac{\gamma}{s-1},$$

and we have

Lemma 2. Let $\mathcal{F}_s$ be a real algebraic number field. Then for any given real number γ, there exists a unit $\eta \in \mathcal{F}_s$ such that

$$c^{-1}e^{\gamma} \le \eta \le ce^{\gamma} \quad \text{and} \quad c^{-1}e^{-\gamma/(s-1)} \le |\eta^{(i)}| \le ce^{-\gamma/(s-1)}, \quad 2 \le i \le s, \tag{17}$$

where $c = c(\mathcal{F}_s)$.

Let $\omega_1, \ldots, \omega_s$ be an integral basis of $\mathcal{F}_s$. Then it follows from Lemma 2 that there exists a sequence of units η_ℓ $(\ell = 1,2,\ldots)$ such that

$$\eta_\ell \geq \ell, \ |\eta_\ell^{(i)}| \leq c(\mathcal{F}_s)\eta_\ell^{-1/(s-1)}, \quad 2 \leq i \leq s. \tag{18}$$

Set

$$\eta_\ell = \sum_{i=1}^{s} \eta_\ell^{(i)} \quad \text{and} \quad h_j^{(\ell)} = \sum_{i=1}^{s} \eta_\ell^{(i)}\omega_j^{(i)}, \quad 1 \leq j \leq s. \tag{19}$$

Then n_ℓ and $h_j^{(\ell)}$ $(1 \leq j \leq s)$ are rational integers.

<u>Theorem 1</u>. We have the simultaneous rational approximations

$$\left|\frac{h_j^{(\ell)}}{n_\ell} - \omega_j\right| \leq c(\mathcal{F}_s)n_\ell^{-1-(1/(s-1))}, \quad 1 \leq j \leq s, \quad \ell = 1,2,\ldots . \tag{20}$$

<u>Proof</u>. For simplicity, we omit the index ℓ. By (18) and (19) we have

$$h = \eta + O(\eta^{-1/(s-1)}) = \eta(1+O(\eta^{-1-(1/(s-1))}))$$

and

$$h_j = \eta\omega_j + O(\eta^{-1/(s-1)}) = \eta\omega_j(1 + O(\eta^{-1-(1/(s-1))})).$$

Hence the theorem follows.

Theorem 1 is not new. The classical methods can only prove the existence of infinitely many sets $(h_1,\ldots,h_s,n)$ satisfying (20). Our purpose is to suggest a <u>computational</u> method for obtaining $(h_1,\ldots,h_s,n)$ satisfying (20). Our method depends on a sequence of units η_ℓ satisfying (18) which can be obtained if a set of independent units of $\mathcal{F}_s$ is known

<u>Example 1</u>. Let p be a prime number ≥ 5 and $s = (p-1)/2$. Then $\mathcal{R}_s = Q(\cos\frac{2\pi}{p})$ is a real cyclotomic field of degree s. It has a set of integral basis

$$\omega_\ell = 2\cos\frac{2\pi\ell}{p}, \quad 1 \leq \ell \leq s$$

and a set of independent units

$$\varepsilon_j = \sin\frac{\pi}{p}g^{j+1}/\sin\frac{\pi}{p}g^j, \quad 1 \leq j \leq s-1,$$

where g is a primitive root mod p. Hence we can obtain a set of units satisfying (18) and consequently a sequence of sets of integers $(h_1^{(\ell)},\dots,h_s^{(\ell)},n_\ell)$ $(\ell = 1,2,\dots)$ satisfying

$$\left|\frac{h_j^{(\ell)}}{n_\ell} - 2\cos\frac{2\pi j}{p}\right| = O(n_\ell^{-1-(1/(s-1))}), \quad 1 \le j \le s-1, \quad \ell = 1,2,\dots .$$

We can prove by means of Schmidt's theorem [1] the following.

Theorem 2. Let $\mathcal{F}_s$ be a real algebraic number field of degree s. Let $\omega_1,\dots,\omega_s$ be an integral basis of $\mathcal{F}_s$. Then the set

$$(\{\omega_1 k\},\dots,\{\omega_{s-1}k\}), \quad 1 \le k \le n \tag{21}$$

has discrepancy $D(n) = O(n^{-1+\varepsilon})$.

Theorem 3. Let $(h_1,\dots,h_{s-1},n)$ be a set of integers satisfying (20). Then the set

$$\left(\frac{k}{n},\left\{\frac{h_1 k}{n}\right\},\dots,\left\{\frac{h_{s-1}k}{n}\right\}\right), \quad 1 \le k \le n \tag{22}$$

has discrepancy $D(n) = O(n^{-1/2-(1/2(s-1))+\varepsilon})$.

We may also start from a single algebraic integer or a unit of an algebraic number field $Q(\alpha)$ by setting $\eta_\ell = \alpha^\ell$. In particular, we may take α to be a PV (Pisot-Vijayaraghavan) number, i.e., α is an algebraic number such that

$$\alpha > 1 \quad \text{and} \quad |\alpha^{(2)}| \le |\alpha^{(3)}| \le \dots |\alpha^{(s)}| < 1. \tag{23}$$

Suppose that α is a PV number and let

$$\rho = -\log|\alpha^{(s)}|/\log\alpha. \tag{24}$$

Then

$$n_\ell = \sum_{i=1}^{s} \alpha^{(i)\ell}$$

is a rational integer and

$$\frac{n_{\ell+k}}{n_\ell} = \frac{\alpha^{\ell+k}(1+O(\alpha^{-\ell}|\alpha^{(s)}|^\ell)}{\alpha^\ell(1+O(\alpha^{-\ell}|\alpha^{(s)}|^\ell)}.$$

Therefore we have

Theorem 4. Let α be a PV number of degree s. Then

$$\left|\frac{n_{\ell+k}}{n_\ell} - \alpha^k\right| = O(n_\ell^{-1-\rho}), \quad 1 \le k \le s-1. \tag{25}$$

If α satisfies the irreducible equation

$$f(x) = x^s - a_{s-1}x^{s-1} - \dots - a_1x - a_0 = 0,$$

then n_ℓ can be evaluated by the Newton's formula of symmetric functions

$$n_0 = s,\ n_1 = a_{s-1}, \dots, n_{s-1} = a_{s-1}n_{s-2} + \dots + a_1n_0$$

and

$$n_\ell = a_{s-1}n_{\ell-1} + a_{s-2}n_{\ell-2} + \dots + a_1n_{\ell-s+1} + a_0n_{\ell-s} \quad (\ell \ge s).$$

Since $\rho \le \frac{1}{s-1}$, it follows that except α is a quadratic irrational or α is a cubic algebraic number such that $\alpha^{(2)}$ and $\alpha^{(3)}$ are complex conjugates, this method is less precise than the prevous one (Minkowski [1]). However the calculation of n_ℓ's is much simpler because they are satisfied by a recurrent formula.

Example 2. Let (F_n) be a sequence of integers defined by

$$F_0 = F_1 = \dots = F_{s-2} = 0,\ F_{s-1} = 1,\ F_{n+s} = F_{n+s-1} + \dots + F_{n+1} + F_n \quad (n \ge 0).$$

Then we have

Theorem 5. We have

$$\left|\frac{F_{n+k}}{F_n} - \eta^k\right| = O(F_n^{-1-(1/2^s\log 2)-(1/2^{2s+1})}), \quad 1 \le k \le s-1, \tag{26}$$

where η is the real solutions of $x^s - x^{s-1} - \dots - x - 1 = 0$.

Theorem 6. The set

$$\left(\frac{k}{F_n}, \left\{\frac{F_{n+1}k}{F_n}\right\}, \dots, \left\{\frac{F_{n+s-1}k}{F_n}\right\}\right), \quad 1 \le k \le F_n \tag{27}$$

has discrepancy $D(F_n) = O(F_n^{-1/2-(1/2^{s+1}\log 2)-(1/2^{2s+3})})$.

§3. Numerical error of quadrature formula.

By the combination of (7) and Theorems 2, 3, and 6, we have the error estimations of the corresponding quadrature formulas. However the error terms in each formula are all ineffective, i.e., the constants implicit in the symbol "O" in Theorems 2, 3, and 6 are ineffective. Sometimes the error term is effective, for example, the constant in (9), but it is too large and is useless in practice. So we shall illustrate a method for estimating the numerical error of quadrature formulas. Suppose that

$$\frac{\partial^{2s} f}{\partial x_1^2 \ldots \partial x_s^2}$$

and its lower derivatives are periodic functions of s variables each with period 1 and are bounded by C. Then f has absolute Fourier expansion

$$f(\underline{\underline{x}}) = \sum_{m_1=-\infty}^{\infty} \cdots \sum_{m_s=-\infty}^{\infty} C(m_1, \ldots, m_s) \exp(2\pi i(m_1 x_1 + \ldots + m_s x_s))$$

$$= \sum C(\underline{\underline{m}}) \exp(2\pi i(\underline{\underline{m}}, \underline{\underline{x}})),$$

where the Fourier coefficient $C(\underline{\underline{m}})$ satisfies

$$|C(\underline{\underline{m}})| \le C \|\underline{\underline{m}}\|^{-2} \tag{28}$$

in which $\|\underline{\underline{m}}\| = \bar{m}_1 \ldots \bar{m}_s$ and $\bar{n} = \max(1, |n|)$. The class of these functions is denoted by $E_s(C)$. Since

$$\frac{1}{m} \sum_{k=0}^{m-1} \exp(2\pi i n k/m) = \begin{cases} 1, & \text{if } m|n, \\ 0, & \text{otherwise} \end{cases}$$

and

$$C(\underline{\underline{0}}) = \int_{G_s} f(\underline{\underline{x}}) d\underline{\underline{x}},$$

we have for any integral vector $\underline{\underline{a}}$

$$\frac{1}{n} \sum_{k=1}^{n} f\left(\frac{\underline{\underline{a}}k}{n}\right) = \frac{1}{n} \sum_{k=1}^{n} \sum_{\underline{\underline{m}}} C(\underline{\underline{m}}) \exp(2\pi i(\underline{\underline{m}}, \underline{\underline{a}})k/n)$$

$$= \sum_{\underline{\underline{m}}} C(\underline{\underline{m}}) \frac{1}{n} \sum_{k=1}^{n} \exp(2\pi i(\underline{\underline{a}}, \underline{\underline{m}})k/n)$$

$$= \sum_{(\underline{a},\underline{m})\equiv 0 \pmod n} C(\underline{m}) = C(\underline{0}) + \sum_{(\underline{a},\underline{m})\equiv 0 \pmod n}' C(\underline{m}),$$

and so by (28),

$$\sup_{f\in E_s(C)} \left| \int_{G_s} f(\underline{x})d\underline{x} - \frac{1}{n}\sum_{k=1}^{n} f\left(\frac{\underline{a}k}{n}\right)\right| \le C\Omega(\underline{a},n),$$

where

$$\Omega(\underline{a},n) = \sum_{(\underline{a},\underline{m})\equiv 0 \pmod n}' \|\underline{m}\|^{-2} \tag{29}$$

in which $\sum'$ denotes a sum by omitting the term $\underline{m} = \underline{0}$.

This is another approach for error estimation of quadrature formula. And we must note that any integral of a non-periodic function can be reduced to an integral of a periodic function.

Example 1. Let

$$\phi_1(\underline{x}) = \frac{1}{2}(f(\underline{x})+f(\underline{x}_1(1-x_1)),\dots,$$

$$\phi_s(\underline{x}) = \frac{1}{2}(\phi_{s-1}(\underline{x})+\phi_{s-1}(\underline{x}_s(1-x_s)),$$

where $\underline{x}_\nu(x)$ is obtained by changing the ν-th component x_ν of $\underline{x}$ by x, i.e., $\underline{x}_\nu(x) = (x_1,\dots,x_{\nu-1},x,x_{\nu+1},\dots,x_s)$. It is clear that $\phi_s(\underline{x})$ is periodic and

$$\int_{G_s} f(\underline{x})d\underline{x} = \int_{G_s} \phi_s(\underline{x})d\underline{x}.$$

Example 2. Let

$$\phi(\underline{x}) = \left(\frac{\pi}{2}\right)^s f\left(\left(\sin\frac{\pi x_1}{2}\right)^2,\dots,\left(\sin\frac{\pi x_s}{2}\right)^2\right)\sin \pi x_1\dots\sin \pi x_s.$$

Then $\phi(\underline{x})$ is a periodic function and

$$\int_{G_s} f(\underline{x})d\underline{x} = \int_{G_s} \phi(\underline{x})d\underline{x}.$$

Theorem 7. We have

$$\Omega(\underline{\underline{a}}, n) = \begin{cases} \frac{1}{n}(1+\frac{\pi^2}{3})^s + \frac{2}{n} \sum_{k=1}^{\frac{n-1}{2}} \prod_{\nu=1}^{s} (1+2\pi^2 B_2(\{\frac{ka_\nu}{n}\})), & \text{if } 2\nmid n \\ \frac{1}{n}(1+\frac{\pi^2}{3})^s + \frac{1}{n}(1-\frac{\pi^2}{6})^\mu (1+\frac{\pi^2}{3})^{s-\mu} + \frac{2}{\pi} \sum_{k=1}^{\frac{n}{2}-1} \prod_{\nu=1}^{s} (1+2\pi^2 B_2(\{\frac{ka_\nu}{n}\})), & \text{if } 2 \mid n, \end{cases} \tag{30}$$

where μ denotes the number of odd integers of a_ν $(1 \le \nu \le s)$ and $B_2(x) = x^2 - x + \frac{1}{6}$ denotes the Bernoulli polynomial.

Proof. Since

$$\Omega(\underline{\underline{a}}, n) = \frac{1}{n} \sum_{k=1}^{n} \sum{}' \exp(2\pi i(\underline{\underline{a}}, \underline{\underline{m}})k/n)\|\underline{\underline{m}}\|^{-2}$$

$$= \frac{1}{n} \sum_{k=1}^{n} \prod_{\nu=1}^{s} \left[\sum_{m=-\infty}^{\infty} \exp(2\pi i a_\nu m_\nu k/n) \bar{m}_\nu^{-2} \right] - 1$$

$$= \frac{1}{n} \sum_{k=1}^{n} \prod_{\nu=1}^{s} (1+2\pi^2 B_2(\{\frac{ka_\nu}{n}\})) - 1,$$

the theorem follows.

For given $\underline{\underline{a}}, n$, the corresponding $\Omega(\underline{\underline{a}}, n)$ can be obtained by the Theorem 7. Korobov suggested to take $\underline{\underline{a}} = (1, a, \ldots, a^{s-1})$ for $n = p$, and then the values of $\Omega(\underline{\underline{a}}, p)$ $(1 \le a < p)$ are compared of which the vector $\underline{\underline{a}}$ such that $\Omega(\underline{\underline{a}}, p)$ takes minimum is a set of optimal coefficients. Hence there requires $O(p^2)$ elementary operations for obtaining a set of optimal coefficients modulo p. The methods of diophantine approximations require only $O(\log n)$ operations for obtaining $\underline{\underline{a}}(n)$. A table containing $(\underline{\underline{a}}, n)$ was compiled which is beneficial for practical use (see Hua and Wang [6]).

§4. Other applications.

1. Interpolation. For any given periodic function f of s variables, we want to find a trigonometrical polynomial P_f to approximate f with the error term that its principal order is independent of s.

Suppose that $f(\underline{\underline{x}}) \in E_s(C)$. Then $f(\underline{\underline{x}})$ has Fourier expansion

$$f(\underline{\underline{x}}) = \sum C(\underline{\underline{m}}) \exp(2\pi i(\underline{\underline{m}}, \underline{\underline{x}})),$$

where

$$C(\underline{m}) = \int_{G_s} f(\underline{x})\exp(-2\pi i(\underline{m},\underline{x}))d\underline{x}.$$

Let

$$P_f(\underline{x}) = \frac{1}{n}\sum_{k=1}^{n} f\left(\frac{\underline{a}k}{n}\right)\sum_{\|\underline{m}\|<N}\exp\left(2\pi i\left(\underline{m},\underline{x}-\frac{\underline{a}k}{n}\right)\right).$$

Then

$$\Delta = \sup_{f\in E_s(C)} |f(\underline{x}) - P_f(\underline{x})| \le \Sigma_1 + \Sigma_2,$$

where

$$\Sigma_1 = \sup_{f\in E_s(C)} \sum_{\|\underline{m}\|<N} \left|C(\underline{m}) - \frac{1}{n}\sum_{k=1}^{n} f\left(\frac{\underline{a}k}{n}\right)\exp\left(2\pi i\left(\underline{n},\underline{x}-\frac{\underline{a}k}{n}\right)\right)\right|$$

and

$$\Sigma_2 = \sup_{f\in E_s(C)} \sum_{\|\underline{m}\|\ge N} |C(\underline{m})|.$$

Since the number of solutions of $\|\underline{m}\| = M$ $(M \ne 0)$ is $O(M^{\varepsilon})$, we have

$$\Sigma_2 \le \sum_{\|\underline{m}\|\ge N} C\|\underline{m}\|^{-2} = O\left(C\sum_{m=N}^{\infty} m^{-2+\varepsilon}\right) = O(CN^{-1+\varepsilon}).$$

Since

$$\frac{1}{n}\sum_{k=1}^{n} f\left(\frac{\underline{a}k}{n}\right)\exp\left(-2\pi i\left(\underline{m},\frac{\underline{a}k}{n}\right)\right) = \frac{1}{n}\sum_{k=1}^{n}\sum_{\underline{\ell}} C(\underline{\ell})\exp(2\pi i(\underline{\ell}-\underline{m},\underline{a})k/n)$$

$$= \sum_{(\underline{a},\underline{\ell}-\underline{m})\equiv 0 \pmod n}^{n} C(\underline{\ell}) = \sum_{(\underline{a},\underline{r})\equiv 0 \pmod n} C(\underline{r}+\underline{m}),$$

we have

$$\left|C(\underline{m}) - \frac{1}{n}\sum_{k=1}^{n} f\left(\frac{\underline{a}k}{n}\right)\exp\left(-2\pi i\left(\underline{m},\frac{\underline{a}k}{n}\right)\right)\right| \le C \sum_{(\underline{a},\underline{r})\equiv 0 \pmod n}{}' \|\underline{m}+\underline{r}\|^{-2}.$$

Since

$$\|\underline{r}\| \le 2^s\|\underline{m}\|\,\|\underline{r}+\underline{m}\|,$$

we have

$$\Sigma_1 \le CN^2 \sum_{(\underline{\underline{a}},\underline{\underline{r}})\equiv 0(\mathrm{mod}\ n)}{}' \|\underline{\underline{r}}\|^{-2} = CN^2\Omega(\underline{\underline{a}},n),$$

and so

$$\sup_{f\in E_s(C)} |f(\underline{\underline{x}}) - P_f(\underline{\underline{x}})| = O(CN^{-1+\varepsilon} + CN^2\Omega(\underline{\underline{a}},n)),$$

where $\Omega(\underline{\underline{a}},n)$ is given in (30). Therefore the error estimation of quadrature formula can be used to estimate the Δ. We can choose $n, \underline{\underline{a}}(n)$ and $N = N(n)$ such that the principal order of Δ is independent of s.

2. Fredholm integral equation of second type. We use the capital Latin letter to denote the s-dimensional vector. Consider the integral equation

$$\phi(P) = \lambda \int_{G_s} K(P,Q)\phi(Q)dQ + f(P),$$

where $f \in E_s(C)$ and $K \in E_{2s}(C)$. Denote $M_k = \frac{\underline{\underline{a}}k}{n}$ $(1 \le k \le n)$. Then we have

$$\int_{G_s} K(P,Q)\phi(Q)dQ = \frac{1}{n}\sum_{k=1}^{n} K(P,M_k)\phi(M_k) + O(\Omega(\underline{\underline{a}},n)).$$

Let $\tilde{\phi}(M_k)$ $(1 \le k \le n)$ be the solution of the system of linear equations

$$\tilde{\phi}(M_j) = \frac{\lambda}{n}\sum_{k=1}^{n} K(M_j,M_k)\tilde{\phi}(M_k) + f(M_j), \quad 1 \le j \le n.$$

Then we can prove that

$$\phi(P) = f(P) + \frac{\lambda}{n}\sum_{k=1}^{n} K(P,M_k)\tilde{\phi}(M_k) + O(\Omega(\underline{\underline{a}},n)),$$

where the constant implicit in the symbol "O" depends only on $f, K,$ and λ.

3. Volterra integral equation of second type. Consider the integral equation

$$\phi(x) = \int_0^x K(x,y)\phi(y)dy + f(x),$$

where $f \in E_1(C)$ and $K \in E_2(c)$. The solution of the equation is given by the Neumann series

$$\phi(x) = f(x) + \sum_{\nu=1}^{\infty} \phi_\nu(x)$$

where

$$\phi_\nu(x) = \int_0^x \int_0^{x_1} \cdots \int_0^{x_{\nu-1}} R_\nu dx_1 \ldots dx_\nu$$

and

$$R_\nu = R_\nu(x, x_1, \ldots, x_\nu) = K(x, x_1)K(x_1, x_2) \ldots K(x_{\nu-1}, x_\nu)f(x_\nu).$$

Since there exists a trigonometrical polynomial P_ν such that

$$|R_\nu - P_\nu| = O(N^{-1+\varepsilon} + N^2(\Omega(\underline{\underline{a}}, n))),$$

we have

$$|\phi_\nu(x) - \int_0^x \cdots \int_0^{x_{\nu-1}} P_\nu dx_1 \ldots dx_\nu| \le \int_0^x \cdots \int_0^{x_{\nu-1}} |R_\nu - P_\nu| dx_1 \ldots dx_{\nu-1}$$

$$= O(\frac{1}{\nu!}(N^{-1+\varepsilon} + N^2\Omega(\underline{\underline{a}}, n))).$$

Dividing the Neumann series into two parts

$$\sum_{n=1}^{\infty} \phi_n(x) = \sum_{\nu \le M}^{\infty} \phi_\nu(x) + \sum_{\nu > M}^{\infty} \phi_\nu(x).$$

The latter series can be evaluated by the trivial estimation

$$\sum_{\nu > M}^{\infty} \phi_\nu(x) = O\left[\sum_{\nu > M}^{\infty} \frac{1}{\nu!}\right] = O(M!^{-1}).$$

Choose suitable M. We have

$$\phi(x) = f(x) + \sum_{\nu=1}^{M} \int_0^x \cdots \int_0^{x_{\nu-1}} P_{\nu-1} dx_1 \ldots dx_\nu + O(N^{-1+\varepsilon} + N^2\Omega(\underline{\underline{a}}, n)).$$

4. We may consider the problem of approximate solution of the least eigenvalue and its corresponding eigenfunction of the homogeneous Fredholm

integral equation

$$\phi(x) = \lambda \int_0^1 K(x,y)\phi(y)dy,$$

where $K(x,y)$ satisfies certain conditions.

5. Consider the Cauchy problem of the equation

$$\frac{\partial u}{\partial t} = \left(\frac{\partial^2}{\partial x_1^2} + \dots + \frac{\partial^2}{\partial x_s^2}\right)u, \quad 0 \le t \le T, \ -\infty < x_\nu < \infty \quad (1 \le \nu \le s),$$

where the initial condition is

$$u(0,\underline{\underline{x}}) = f(\underline{\underline{x}}) \in E_s(C).$$

We may prove by the use of the result of interpolation that

$$u(t,\underline{\underline{x}}) = \sum_{\|\underline{\underline{m}}\|<N} \frac{1}{n} \sum_{k=1}^{n} f\left(\frac{\underline{\underline{a}}k}{n}\right)\exp(-2\pi i(\underline{\underline{a}},\underline{\underline{m}})k/m - 4\pi^2(\underline{\underline{m}},\underline{\underline{m}}) + 2\pi i(\underline{\underline{m}},\underline{\underline{x}}))$$

$$+ O(N^{-1+\varepsilon} + N^2(\Omega(\underline{\underline{a}},n))).$$

6. Consider the Dirichlet problem of the equation

$$\left(\frac{\partial^2}{\partial x_1^2} + \dots + \frac{\partial^2}{\partial x_s^2}\right)u = f,$$

where $f \in E_s(C)$ and f is an odd function with respect to each variable and u satisfies the boundary condition that $u(\underline{\underline{x}}) = 0$ on G_s. We can obtain an approximate solution of $u(\underline{\underline{x}})$:

$$u^*(\underline{\underline{x}}) = \sum_{k=1}^{n} f\left(\frac{\underline{\underline{a}}k}{n}\right)\psi_k(\underline{\underline{x}}).$$

where

$$\psi_k(\underline{\underline{x}}) = -\frac{1}{4\pi^2 n} \sum_{\|\underline{\underline{m}}\|<N}{}' \exp\left(2\pi i\left(\underline{\underline{m}},\underline{\underline{x}} - \frac{\underline{\underline{a}}k}{n}\right)\right)(\underline{\underline{m}},\underline{\underline{m}})^{-1}, \quad 1 \le k \le n.$$

The difference between $u(\underline{\underline{x}})$ and $u^*(\underline{\underline{x}})$ is independent of s in its principal term (see Hua and Wang [4,6], Wang [2,3,4,6]).

§5. Some conjectures.

For further studies, we would like to pose the following conjectures:

1. There exists a set of integers (n_ℓ) $(1 < n_1 < n_2 < \dots)$ and integral vectors $\underline{a} = \underline{a}(n_\ell) = (a_1, \dots, a_s)$ $(\ell = 1, 2, \dots)$ such that

$$W(n_\ell, \underline{a}) = \sup_{f \in E_s(C)} \left| \int_{G_s} f(\underline{x}) d\underline{x} - \frac{1}{n_\ell} \sum_{k=1}^{n_\ell} f\left(\frac{\underline{a}k}{n_\ell}\right) \right| \le \frac{Cc(s)(\log n_\ell)^{s-1}}{n_\ell^2}. \tag{31}$$

Hua and Wang [1] and Bahvalov [1] proved independently for $s = 2$, $n_\ell = F_{\ell+2}$, $\underline{a} = (1, F_{\ell+1})$ $(\ell = 1, 2, \dots)$. Korobov [1] established that there exists a vector $\underline{a} = \underline{a}(p)$ for each prime p such that $W(p, \underline{a}) \le C \cdot c(s) \frac{(\log p)^{2s}}{p^2}$, and it was improved to $W(p, \underline{a}) \le \frac{C \cdot c(s)(\log p)^{2(s-1)}}{p^2}$ by Bahvalov [1]. Sarygin [1] proved that $W(n, \underline{a}) \ge Cc(s) \frac{(\log n)^{s-1}}{n^2}$ holds for any integer $n \ge 2$ and integral vector $\underline{a} = \underline{a}(n)$. Bahvalov [1] also proved that (31) can be derived from the following conjecture.

2. There exists a set of integers (n_ℓ) and integral vectors $\underline{a} = \underline{a}(n_\ell)$ $(\ell = 1, 2, \dots)$ such that any nontrivial solution of the congruence

$$(\underline{a}, \underline{m}) \equiv 0 \pmod{n_\ell}$$

satisfies

$$\|\underline{m}\| > c(s) n_\ell.$$

3. There exists a set of integers (n_ℓ) and integral vectors $\underline{a} = \underline{a}(n_\ell)$ such that the set $\left(\left\{\frac{a_1 k}{n_\ell}, \dots, \frac{a_s k}{n_\ell}\right\}\right)$ $(1 \le k \le n_\ell)$ has discrepancy

$$D(n_\ell) \le c(s) \frac{(\log n_\ell)^{s-1}}{n_\ell}.$$

Zaremba [1] established $D(F_{\ell+2}) < cF_{\ell+2}^{-1} \log F_{\ell+2}$ for $s = 2$ and $\underline{a} = (1, F_{\ell+2})$ (see Wang [5]).

References

N.S. Bahvalov

[1] On approximate evaluation of multiple integrals, Ves. Univ. of Moscow, 1959, 3-18.

S. Haber

[1] Experiments on optimal coefficients, see "Applications of number theory to numerical analysis," edited by S. Zaremba, Acad. Press, 1972, 11-18.

E. Hlawka

[1] Funktionen von beschränkter Variation in der Theorie Gleichverteilung, Ann. di Math. Pure Appl. 4, 1961, 325-335.

[2] Uniform distribution modulo 1 and numerical integration, Comp. Math. 16, 1964, 95-105.

Hua Lookeng and Wang Yuan

[1] Remarks concerning numerical integration, Sci. Rec., New Ser, 4 1960, 8-11.

[2] On diophantine approximations and numerical integrations I, Sci. Sin, 6, 13, 1964, 1007-1008; II, Sci. Sin, 6, 13, 1964, 1009-1010.

[3] On numerical integration of periodic functions of several variables, Sci. Sin., 14, 7, 1965, 964-978.

[4] On uniform distribution and numerical analysis (Number theoretic method) I, Sci. Sin., 4, 16, 1973, 483-505; II, Sci. Sin. 3, 17, 1974, 331-348; III, Sci. Sin., 2, 18, 1975, 184-198.

[5] A note on simultaneous diophantine approximations to algebraic integers, Sci. Sin., 5, 20, 1977, 563-567.

[6] Applications of number theory to numerical analysis, Springer Verlag (Heidelberg) and Science Press (Beijing), 1981.

N.M. Korobov

[1] On approximate evaluation of multiple integrals, Dokl. Akad. Hayk SSSR, 124, 6, 1959, 1207-1210.

[2] Properties and calculation of optimal coefficients, Dokl. Akad. Hayk SSSR, 132, 1960, 1009-1012.

[3] Number theoretic method in approximate analysis, Fizmat, Moscow, 1963.

H. Minkowski

[1] Über periodische Approximationen algebraischer Zahlen, Acta Math., 26, 1902, 333-351.

H. Niederreiter

[1] Existence of good lattice points in the sense of Hlawka, Monatsh. Math. 86, 1978, 203-219.

K.F. Roth

[1] On irregularities distribution, Mathematica, 2, 1954, 73-79.

I.F. Sarygin

[1] A lower estimation for the error of quadrature formulas for certain classes of functions, Z. Vycisl. Mat. i Mat. Fiz., 3, 1963, 370-376.

W.M. Schmidt

[1] Simultaneous approximation to algebraic numbers by rationals, Acta Math. 125, 1970, 189-201.

I.M. Sobol

[1] An exact estimate of the error in multidimensional quadrature formulas for functions of the classes $\tilde{W}_1$ and $\tilde{H}_1$, Z. Zycisl. Mat. i. Mat. Fiz., 1, 1961, 208-216.

Wang Yuan

[1] A note on interpolation of a certain class of functions, Sci. Sin., 10, 1960, 632-636.

[2] On numerical integration and its applications (Number theoretic method), Shuxue Jihzhan, 5, 1962, 1-44.

[3] Remarks on the interpolation of a certain class of functions, Sci. Sin., 14, 1965, 629-631.

[4] On interpolation of a certain class of functions, Kexue Tongbao, 9, 1966, 387-389.

[5] On diophantine approximation and approximate analysis I, Acta Math. Sin., 25, 1982, 248-256; II, Acta Math. Sin., 25, 1982, 323-332.

[6] A note on the approximate solution of the Cauchy problem by number theoretic nets, Chinese Ann. Math., 3, 1982, 451-456.

Wang Yuan, Xu Guangshan and Zhang Rongxiao,

[1] On number theoretic method of numerical integration in multi-dimensional space I, Acta Math. Appl. Sin., 2, 1978, 106-114; II, Acta Math. Appl. Sin., 5, 1982, 412-417.

H. Weyl

[1] Über die Gleichverteilung der Zahlen mod. Eins, Math. Ann., 77, 1916, 313-352.

S.K. Zaremba

[1] Good lattice points, discrepancy and numerical integration, Ann. Mat. Pure Appl. 73, 1966, 293-317.

Institute of Mathematics
Academia Sinica
Beijing

Contemporary Mathematics
Volume **77**, 1988

DIOPHANTINE EQUATIONS AND DIOPHANTINE INEQUALITIES IN ALGEBRAIC NUMBER FIELDS

Wang Yuan

§1. Introduction. Given a vector with complex coefficients $\underline{\underline{x}} = (x_1,\ldots,x_s)$ put

$$|\underline{\underline{x}}| = \max|x_i|$$

and, give a form (i.e., a homogeneous polynomial) F, let

$$|F|$$

be the maximum absolute value of its coefficients. With every form F of degree k there is associated a form

$$\hat{F}(\underline{\underline{x}}_1,\ldots,\underline{\underline{x}}_k)$$

which is linear in each vector $\underline{\underline{x}}_i$ $(1 \le i \le k)$ and symmetric in the k vectors $\underline{\underline{x}}_1,\ldots,\underline{\underline{x}}_k$ and such that

$$F(\underline{\underline{x}}) = \hat{F}(\underline{\underline{x}},\ldots,\underline{\underline{x}}).$$

Schmidt [4] in 1980 proved

Theorem A (Schmidt). Given $h \ge 1$, $m \ge 1$ and odd numbers $k_1,\ldots,k_h$, and given a positive number E, however, large, there is a constant

$$c_1 = c_1(k_1,\ldots,k_h;m,E)$$

as follows. If $M \ge 1$ is real and if $F_1,\ldots,F_h$ are forms with real coefficients of respective degrees $k_1,\ldots,k_h$ in $\underline{\underline{x}} = (x_1,\ldots,x_s)$ where $s \ge c_1$, then there are m linearly independent integer points $\underline{\underline{x}}(1),\ldots,\underline{\underline{x}}(m)$ in $\mathbb{Z}^s$ with

$$|\underline{\underline{x}}(i)| \le M, \quad 1 \le i \le m$$

and

$$|\hat{F}_j(\underline{\underline{x}}(i_1),\dots,\underline{\underline{x}}(i_{k_j}))| \ll M^{-E}|F_j|, \quad 1 \le j \le h, \quad 1 \le i_1,\dots,i_{k_j} \le m.$$

If the coefficients of all forms are rational integers, he derives from Theorem A the following.

Theorem B (Schmidt). Given $h \ge 1$, $m \ge 1$ and odd numbers $k_1,\dots,k_h$, and give $\varepsilon > 0$, however small, there is a constant $c_2 = c_2(k_1,\dots,k_h;m,\varepsilon)$ such that if $G_1,\dots,G_h$ are forms of respective degrees $k_1,\dots,k_h$ with integer coefficients in $\underline{\underline{x}} = (x_1,\dots,x_s)$ where $s \ge c_2$, then $G_1,\dots,G_h$ vanish on an m-dimensional subspace which is spanned by integer points $\underline{\underline{x}}(1),\dots,\underline{\underline{x}}(m)$ having

$$|\underline{\underline{x}}(i)| \ll G^{\varepsilon}, \quad 1 \le i \le m,$$

where $G = \max(1,|G_1|,\dots,|G_h|)$.

The first step of the proof of Theorem A is to prove a result on additive equations, namely

Proposition A (Schmidt [3]). Suppose that $\varepsilon > 0$ and that

$$D(\underline{\underline{x}}) = d_1x_1^k + \dots + d_sx_s^k$$

is an additive form of odd degree k with $d_i \in \mathbb{Z}$ $(1 \le i \le s)$ and with $s \ge c_3(k,\varepsilon)$. Then there is a nonzero integer point $\underline{\underline{x}}$ in $\mathbb{Z}^s$ with

$$D(\underline{\underline{x}}) = 0 \quad \text{and} \quad |\underline{\underline{x}}| \le \max(1,|D|^{\varepsilon}).$$

The next step is to prove a special case of Theorem A, namely when there is only one additive form. And the final step is to prove the theorem by an inductive argument.

In the proof of Proposition A, the Hardy-Littlewood circle method is used. It was Siegel [5,6] who succeeded in dealing with Waring's problem in an arbitrary algebraic number field by his generalized circle method, and he obtained the result corresponding the Hardy-Littlewood's estimation on $G(k)$.

By the combination of the methods of Schmidt and Siegel, we can generalize Proposition A to an arbitrary algebraic number field. In particular, if the coefficients of the additive equations belong to a totally complex algebraic number field, i.e., a field with no isomorphic embedding into the reals, the result holds for equations of arbitrary degree k (see Proposition 1).

Let K be an algebraic number field of degree n. Let $K^{(p)}$ $(1 \le p \le r_1)$ be the real conjugates of K and let $K^{(q)}$ and $K^{(q+r_2)}$ $(r_1+1 \le q \le$

r_1+r_2) denote the complex conjugates of K, where $r_1 + 2r_2 = n$. Throughout this paper, the indices p and q are over the sets of integers cited above. For $\gamma \in K$, we denote by $\gamma^{(i)}$ $(1 \le i \le n)$ the conjugates of γ and $N(\gamma) = \prod_{i=1}^{n} \gamma^{(i)}$ the norm of γ. Let γ_j $(1 \le j \le n)$ be numbers of K and x_j $(1 \le j \le n)$ be real numbers. We set $\xi = \sum_{j=1}^{n} x_j\gamma_j$ and define $\xi^{(i)} = \sum_{j=1}^{n} x_j\gamma_j^{(i)}$ $(1 \le i \le n)$. Note that ξ may not belong to K and $\xi^{(i)}$ not a conjugate of ξ. We use the notations

$$T(\xi) = \sum_{j=1}^{n} \xi^{(i)}, \quad \exp(x) = e^x, \quad E(\xi) = \exp(2\pi i T(\xi)), \quad \|\xi\| = \max|\xi^{(i)}|$$

and

$$\|\underline{\underline{\lambda}}\| = \max\|\lambda_i\|$$

for a vector $\underline{\underline{\lambda}} = (\lambda_1,\lambda_2,\ldots,\lambda_s)$.

Our main result is

Theorem 1 [9]. Let K be a totally complex algebraic number field of degree 2r. Given positive integers h, m and $k_1,\ldots,k_h$, and given a positive number E, however large, there is a constant

$$c_4 = c_4(k_1,\ldots,k_h;r,m,E)$$

as follows. If $M \ge 1$ is real and $F_1,\ldots,F_h$ are forms with complex coefficients of respective degrees $k_1,\ldots,k_h$ in $\underline{\underline{\lambda}} = (\lambda_1,\ldots,\lambda_s)$ where $s \ge c_4$, then there are m independent integer points $\underline{\underline{\lambda}}(1),\ldots,\underline{\underline{\lambda}}(m)$ in K^s with

$$\|\underline{\underline{\lambda}}(i)\| \le M, \quad 1 \le i \le m$$

and

$$|\hat{F}_j(\underline{\underline{\lambda}}(i_1),\ldots,\underline{\underline{\lambda}}(i_{k_j}))| \ll M^{-E}|F_j|, \quad 1 \le j \le h, \quad 1 \le i_1,\ldots,i_{k_j} \le m,$$

here and below the constants implicit in $\ll$ or O may depend on $k,k_1,\ldots,k_h,K,m,E,\varepsilon$, but not on $A,M,F_1,\ldots,F_h$ and G.

In particular, it follows that

$$|F_j(\underline{\underline{\lambda}}(i))| \ll M^{-E}|F_j|, \quad 1 \le j \le h, \quad 1 \le i \le m.$$

Suppose now that $G_1,\ldots,G_h$ are forms of respective degrees $k_1,\ldots,k_h$ with coefficients in integer of K. Let

$$\|G_i\|$$

denote the maximum absolute value of its coefficients and their conjugates. Further, let

$$G = \max(1, \|G_1\|, \ldots, \|G_h\|).$$

Suppose that $s \geq c_4(k_1, \ldots, k_h; r, m, 4rk_1 \ldots k_h \varepsilon^{-1}) = c_5(k_1, \ldots, k_h, r, m, \varepsilon)$, say, where $0 < \varepsilon < 1$. Apply Theorem 1 with $M = M_0 G^{\varepsilon}$, where $M_0 = M_0(k_1, \ldots, k_h; K, m, \varepsilon)$ is to be chosen in a moment. We obtain linearly independent integer points $\underline{\underline{\lambda}}(1), \ldots, \underline{\underline{\lambda}}(m)$ in K^s with

$$\|\underline{\underline{\lambda}}(i)\| \leq M_0 G^{\varepsilon}, \quad 1 \leq i \leq m$$

and

$$|\hat{G}(\underline{\underline{\lambda}}(i_1), \ldots, \underline{\underline{\lambda}}(i_{k_j}))| \ll M_0^{-4rk_1, \ldots, k_h} G^{-4rk_1, \ldots, k_h + 1}.$$

On the other hand $k_j! \hat{G}_j(\underline{\underline{\lambda}}(i_1), \ldots, \underline{\underline{\lambda}}(i_{k_j}))$ is an integer in K. If $\hat{G}_j(\underline{\underline{\lambda}}(i_1), \ldots, \underline{\underline{\lambda}}(i_{k_j})) \neq 0$, then it follows from

$$\|k_j! \hat{G}(\underline{\underline{\lambda}}(i_1), \ldots, \underline{\underline{\lambda}}(i_{k_j}))\| \ll M_0^{k_j} G^{1+k_j}$$

and

$$|N(k_j! \hat{G}_j(\underline{\underline{\lambda}}(i_1), \ldots, \underline{\underline{\lambda}}(i_{k_j}))| \geq 1$$

that

$$|\hat{G}_j(\underline{\underline{\lambda}}(i_1), \ldots, \underline{\underline{\lambda}}(i_{k_j}))| \gg (M_0^{k_j} G^{1+k_j})^{-2r+1} \gg M_0^{-2rk_j} G^{-4rk_j+1}$$

which leads to a contradiction if M_0 is sufficiently large. Therefore

$$\hat{G}_j(\underline{\underline{\lambda}}(i_1), \ldots, \underline{\underline{\lambda}}(i_{k_j})) = 0,$$

and we have the following.

<u>Theorem 2</u>. Given positive integers $k_1, \ldots, k_h$ and m, and given ε, however small, there is a constant $c_5 = c_5(k_1, \ldots, k_h; r, m, \varepsilon)$ such that if $G_1, \ldots, G_h$ are forms of respective degrees $k_1, \ldots, k_h$ in $\underline{\underline{\lambda}} = (\lambda_1, \ldots, \lambda_s)$ with coefficients in integer of K, where $s \geq c_5$, then $G_1, \ldots, G_h$ vanish on an m-dimensional subspace which is spanned by integer points $\underline{\underline{\lambda}}(1), \ldots, \underline{\underline{\lambda}}(m)$ in K^s having

$$\|\underline{\underline{\lambda}}(i)\| \ll G^{\varepsilon}, \quad 1 \le i \le m.$$

Theorem 2 gives an improvement of a result due to Peck [1]. He first established the existence of $\underline{\underline{\lambda}}(i)$ $(1 \le i \le m)$ with $c_6(k_1,\dots,k_h;K,m)$ instead of c_5.

§2. Additive equations. Let K be an algebraic number field of degree n. A number γ of K is called totally nonnegative if $\gamma^{(p)} \ge 0$. Let $\alpha_1,\dots,\alpha_s$ be a set of integers in K. Consider the additive form

$$A(\underline{\underline{a}},\underline{\underline{\lambda}}) = \alpha_1 a_1 \lambda_1^k + \dots + \alpha_s a_s \lambda_s^k.$$

A set of numbers $\underline{\underline{a}},\underline{\underline{\lambda}}$ is called a nontrivial solution of the equation

$$A(\underline{\underline{a}},\underline{\underline{\lambda}}) = 0 \tag{1}$$

if each a_i is 1 or -1, and λ_1 $(1 \le i \le s)$ are totally nonnegative integers of K, not all zero. Write

$$A = \max(1, |\alpha_1\|, \dots, \|\alpha_s\|).$$

Proposition 1. Suppose $s \ge c_7(k,n,\varepsilon)$. Then the equation (1) has a nontrivial solution with

$$\|\underline{\underline{\lambda}}\| \ll A^{\varepsilon}. \tag{2}$$

This proposition was proved in [8]. Now we sketch the proof as follows.

1) One can prove by Siegel's method that if $s \ge c_8(k,n)$, then (1) has a nontrivial solution with

$$\|\underline{\underline{\lambda}}\| \ll A^{c_9(k,n)}. \tag{3}$$

We may suppose clearly that $\alpha_i \ne 0$ $(1 \le i \le s)$ and by (3) that $A \ge c_{10}(k,K,\varepsilon)$. Since

$$1 \le |N(\alpha_j)| = |\alpha_j^{(1)} \dots \alpha_j^{(n)}| \le |\alpha_j^{(i)}| A^{n-1},$$

we have

$$\min_{i,j} |\alpha_j^{(i)}| \ge A^{-n+1}.$$

Therefore, there are not less than $s_1 = s/([n/\varepsilon_1]+1)$ numbers among the

α_i's, say $\alpha_1,\dots,\alpha_{s_1}$, satisfying

$$|\alpha_j^{(1)}/\alpha_\ell^{(1)}| \le A^{\varepsilon_1}, \quad 1 \le j,\ell \le s_1.$$

We may suppose without loss of generality that $\alpha_1,\dots,\alpha_s$ satisfy the above relation. Similarly, we may suppose further that $\alpha_1,\dots,\alpha_s$ satisfy

$$|\alpha_j^{(i)}/\alpha_\ell^{(i)}| \le A^{\varepsilon_1}, \quad 1 \le j,\ell \le s, \quad 1 \le i \le n.$$

Take units $\sigma_1,\dots,\sigma_s$ such that

$$|N(\alpha_j)|^{1/n} \ll |\alpha_j^{(i)}\sigma_j^{(i)k}| \ll |N(\alpha_j)|^{1/n}, \quad 1 \le j \le s, \quad 1 \le i \le n.$$

Let $a^n = A^{n\varepsilon_1}\max|N(\alpha_i)|$ and let p_i be the largest rational integer such that

$$|N(\alpha_i)|p_i^{kn} \le a^n, \quad 1 \le i \le s.$$

Since $A \ge c_{10}$, we have

$$\tfrac{1}{2}a^n \le |N(\alpha_i)|p_i^{kn}, \quad 1 \le i \le s.$$

Set $\alpha_i' = \alpha_i\sigma_i^k p_i^k$ and $\lambda_i = \sigma_1^{-1}\sigma_i p_i\lambda_i'$ $(1 \le i \le s)$. Consider the equation

$$A'(\underline{\underline{a}},\underline{\underline{\lambda}}') = \alpha_1'a_1\lambda_1'^k + \dots + \alpha_s'a_s\lambda_s'^k = \sigma_1^k A(\underline{\underline{a}},\underline{\underline{\lambda}}) = 0. \tag{4}$$

Since $p_i \le A^{2\varepsilon_1/k}$ $(1 \le i \le s)$ and $|\sigma_1^{(i)-1}\sigma_j^{(i)}| \ll A^{2\varepsilon_1/k}$ $(1 \le j \le s,\ 1 \le i \le n)$, it follows that Proposition 1 is true if it holds for special equation (4). So we may suppose that $\alpha_1,\dots,\alpha_s$ satisfy

$$c_{11}a < |\alpha_j^{(i)}| < c_{12}a, \quad 1 \le j \le s, \quad 1 \le i \le n, \tag{5}$$

where $c_{11} = c_{11}(k,K)$, $c_{12} = c_{12}(k,K)$ and $a > c_{10}(k,K,\varepsilon)$.

2) We may prove that Proposition 1 follows by the following.

Proposition 2. Suppose that α_i $(1 \le i \le s)$ satisfy (5). If $s \ge c_{13}(k,n,\varepsilon)$, then either (1) has a nontrivial solution with (2) or there is a nonzero integer χ such that

$$A(\underline{\underline{a}},\underline{\underline{\chi}}) = \chi, \quad \|\underline{\underline{\lambda}}\| \le a^{\varepsilon_2}, \quad \|\chi\| \le a^{\varepsilon_3},$$

where each a_i is 1 or -1, and λ_j $(1 \le j \le s)$ are totally nonnegative integers of K, not all zero, and $\varepsilon_2 = \varepsilon_2(\varepsilon)$, $\varepsilon_3 = \varepsilon_3(\varepsilon)$.

3) Proposition 2 is proved by the Siegel's generalized circle method. Let $\omega_1, \ldots, \omega_n$ be an integral basis of K with $\omega^{(q)} = \bar{\omega}^{(r_1+q)}$ $(1 \le q \le r_2)$ and δ the different of K. Then a basis $\rho_1, \ldots, \rho_n$ of δ^{-1} can be defined by

$$T(\omega_i \delta_j) = \begin{cases} 1, & \text{if } i = j, \\ 0, & \text{if } i \neq j. \end{cases}$$

Let G be the n-dimensional unit cube, $\xi = x_1\rho_1 + \ldots + x_n\rho_n$ and $dx = dx_1 \ldots dx_n$. Then for any integer $\mu = y_1\omega_1 + \ldots + y_n\omega_n$ of K we have

$$\int_G E(\mu\xi)dx = \prod_{j=1}^{n} \int_0^1 \exp(2\pi i y_j x_j)dx_j = \begin{cases} 1, & \text{if } \mu = 0, \\ 0, & \text{if } \mu \neq 0. \end{cases}$$

Let

$$S_i(\xi) = \sum_{\lambda \in P(a^{\varepsilon_2})} E(a_i \alpha_i \xi \lambda^k), \quad 1 \le i \le s,$$

where λ runs over all integers satisfying $\lambda^{(p)} \le a^{\varepsilon_2}$, $|\lambda^{(q)}| \le a^{\varepsilon_2}$. Then for a given vector $\underline{\underline{a}}$, where $a_i = \pm 1$, the number of solutions of $A(\underline{\underline{a}}, \underline{\underline{\lambda}}) = \chi$ with $\lambda_i \in P(a^{\varepsilon_2})$ $(1 \le i \le s)$, $\chi \in P(a^{\varepsilon_3})$ is equal to

$$Z = \int_G \sum_{\chi \in P(a^{\varepsilon_3})} S_1(\xi) \ldots S_s(\xi) E(-\xi\chi)dx.$$

Now we define the basic domains and supplementary domain as follows. For any $\gamma \in K$, we can determine uniquely integral ideals $\mathfrak{a}, \mathfrak{c}$ such that

$$\gamma\delta = \mathfrak{c}/\mathfrak{a}, \quad (\mathfrak{a}, \mathfrak{c}) = 1.$$

If $N(\mathfrak{a}) \le t$, we define basic domain B_γ by

$$\{(x_1, \ldots, x_n) : (x_1, \ldots, x_n) \in G,\ \xi = x_1\rho_1 + \ldots + x_n\rho_n \text{ such that}$$
$$H\|\xi - \gamma_0\| < 1 \text{ for some } \gamma_0 \equiv \gamma \pmod{\delta^{-1}}\}.$$

Choose suitable t and H. We can prove that $B_\gamma \cap B_{\gamma'} = \emptyset$ if $\gamma \neq \gamma'$. Set

$$B = \bigcup_\gamma B_\gamma \quad \text{and} \quad E = G - B.$$

We call E the supplementary domain. Therefore

$$Z = \int_B + \int_E .$$

The integral

$$\int_B = J_0 \mathcal{g}(t, a^{\varepsilon_3}) a^{\varepsilon_2 n(s-k)-n}(1+o(1)) \quad (\mathcal{g}(t, a^{\varepsilon_3}) \gg a^{\varepsilon_3 n})$$

gives the principal term of Z. The integral over E can be treated by the following.

Lemma (Schmidt). Suppose that $|\sum_{\lambda \in P(T)} E(\xi\lambda^k)| \geq C$ and $C \geq T^{n-(1/Q)+\eta}$, where $Q = 2^{k-1}$ and $\eta > 0$. Then there exist integers α and β such that

$$0 < \|\alpha\| \ll \left(\frac{T^n}{C}\right)^Q T^\eta \quad \text{and} \quad \|\alpha\xi - \beta\| \ll \left(\frac{T^n}{C}\right)^Q T^{-k+\eta}$$

(see, e.g., [7]).

Since s is large, we can choose a vector $\underline{\underline{a}}$ such that for any p, the numbers $a_1\alpha_1^{(p)}, \ldots, a_s\alpha_s^{(p)}$ have different signs. Then $J_0 > 0$ and so $Z > 0$. The proposition is proved.

Consider the equation

$$A(\underline{\underline{\lambda}}) = \sum_{i=1}^{s} \alpha_i \lambda_i^k = 0, \tag{6}$$

where $\alpha_1, \ldots, \alpha_s$ are given integers in K. If k is an odd number, then $a_i\lambda_i^k = (a_i\lambda_i)^k$ $(1 \leq i \leq s)$. If $r_1 = 0$, i.e., K is totally complex, then J_0 is always positive. Therefore we have

Proposition 3. Suppose that $s \geq c_{14}(k, n, \varepsilon)$. Then the equation (6) has a solution in integers $\lambda_1, \ldots, \lambda_s$ of K, not all zero, satisfying

$$\|\underline{\underline{\lambda}}\| \ll A^\varepsilon,$$

provided only that k is odd or that K is totally complex.

§3. Additive forms. Let K be a totally complex algebraic number field of degree $2r$. Consider the additive form

$$A(\underline{\underline{\lambda}}) = \alpha_1\lambda_1^k + \ldots + \alpha_s\lambda_s^k,$$

where α_i $(1 \leq i \leq s)$ are given complex numbers.

Proposition 4. Given $k \ge 1$ and E, however large, there is a constant $c_{15}(k,r,E)$ such that if $s \ge c_{15}$, then for real $M \ge 1$ there is a nonzero integer point $\underline{\underline{\lambda}} \in K^s$ with

$$\|\underline{\underline{\lambda}}\| \le M \quad \text{and} \quad |A(\underline{\underline{\lambda}})| \ll M^{-E}|A|.$$

This proposition is a very special case of Theorem 1. On the other hand, Theorem 1 can be deduced by an inductive argument based on this proposition (see [9]).

Now we sketch the proof of Proposition 4. As the case in the proof of Proposition 1, we may assume that the coefficiens of $A(\underline{\underline{\lambda}})$ satisfy

$$0 < \frac{1}{2} \le |\alpha_j| \le 1, \quad 1 \le j \le s.$$

We define for each j,

$$\alpha_j^{(q)} = \alpha_j, \quad \alpha_j^{(q+r)} = \bar{\alpha}_j, \quad 1 \le q \le r.$$

Then α_j has a unique representation

$$\alpha_j^{(i)} = z_1\omega_1^{(i)} + \dots + z_{2r}\omega_{2r}^{(i)}, \quad 1 \le i \le 2r,$$

where $z_\ell \in \mathbb{R}$ $(1 \le \ell \le 2r)$. Set $\xi = x_1\rho_1 + \dots + x_{2r}\rho_{2r}$, $\eta = y_1\omega_1 + \dots + y_{2r}\omega_{2r}$, $dx = dx_1 \dots dx_{2r}$, $dy = dy_1 \dots dy_{2r}$, and define

$$S_i(\xi) = \sum_{\|\lambda\| \le M} E(\alpha_i \xi \lambda^k) \quad \text{and} \quad I_i(\xi) = \int_{\|\eta\| \le M} E(\alpha_i \xi \eta^k)dy, \quad 1 \le i \le s.$$

Let Z be the number of solutions of

$$|A(\underline{\underline{\lambda}})| \ll M^{-E}$$

in integer points $\underline{\underline{\lambda}} \in K^s$ subject to

$$\|\underline{\underline{\lambda}}\| \le M.$$

Expand $\sum_{j=1}^{s} \alpha_j \lambda_j^k$ as

$$\sum_{j=1}^{s} \alpha_j^{(i)} \lambda_j^{(i)k} = A_1\omega_1^{(i)} + \dots + A_{2r}\omega_{2r}^{(i)}, \quad 1 \le i \le 2r.$$

Since for real Q

$$M^E \int_{-\infty}^{\infty} \exp(2\pi i\beta Q)\left(\frac{\sin \beta M^{-E}}{\pi\beta}\right)^2 d\beta = \begin{cases} 1-M^E|Q|, & \text{if } |Q| < M^{-E}, \\ 0, & \text{if } |Q| \ge M^{-E}, \end{cases} \tag{7}$$

we have

$$\begin{aligned} M^{2rE} & \int_{-\infty}^{\infty} \cdots \int \prod_{\ell=1}^{s} S_\ell(\xi) \prod_{j=1}^{2r} \left(\frac{\sin \pi x_j M^{-E}}{\pi x_j}\right)^2 dx \\ &= M^{2rE} \sum_{\|\lambda_1\| \le M} \cdots \sum_{\|\lambda_s\| \le M} \prod_{j=1}^{2r} \int_{-\infty}^{\infty} \exp(2\pi i x_j A_j) \left(\frac{\sin \pi x_j M^{-E}}{\pi x_j}\right)^2 dx_j \\ &= \sum_{\|\lambda_1\| \le M} \cdots \sum_{\substack{\|\lambda_s\| \le M \\ |A_i| < M^{-E}}} (1-|A_1|M^E)\cdots(1-|A_{2r}|M^E). \end{aligned} \tag{8}$$

Let $\eta_j = y_{j1}\omega_1 + \cdots + y_{j,2r}\omega_{2r}$, $dY_j = dy_{j1}\cdots dy_{j,2r}$ $(1 \le j \le s)$ and

$$\sum_{j=1}^{s} \alpha_j^{(i)} \eta_j^{(i)k} = B_1\omega_1^{(i)} + \cdots + B_{2r}\omega_{2r}^{(i)}, \quad 1 \le i \le 2r.$$

Then by (7) we have

$$\begin{aligned} M^{2rE} & \int_{-\infty}^{\infty} \cdots \int \prod_{\ell=1}^{s} I_\ell(\xi) \prod_{j=1}^{2r} \left(\frac{\sin \pi x_j M^{-E}}{\pi x_j}\right)^2 dx \\ &= \int_{\|\eta_1\| \le M} \cdots \int_{\|\eta_s\| \le M} (1-|B_1|M^E)\cdots(1-|B_{2r}|M^E)\, dY_1 \cdots dY_s. \end{aligned} \tag{9}$$

If $|A_i| < M^{-E}$, $1 \le i \le 2r$, then it follows that $A(\underline{\lambda}) \ll M^{-E}$, and thus the right hand side of (8) gives a lower bound for Z. The general idea now will be to show that the right hand side of (9) is large and to show that the left hand sides of (8) and (9) differ little.

Remarks. 1. We have also studied a special type of additive equation

$$\alpha_1\lambda_1^k + \cdots + \alpha_s\lambda_s^k = \beta_1\mu_1^k + \cdots + \beta_s\mu_s^k, \tag{10}$$

where $\alpha_1, \ldots, \alpha_s$, $\beta_1, \ldots, \beta_s$, are $2s$ nonzero totally nonnegative integers of K. And we proved that if $s \ge c_{16}(k, n, \varepsilon)$, then the equation (10) has a nontrivial solution such that

$$\max_{i,j}(N(\lambda_i),N(\mu_j)) \ll \max_{i,j}(N(\alpha_i),N(\beta_j))^{(1/k)+\varepsilon} \tag{11}$$

(see [7]). This gives a generalization of a result due to Schmidt [2]. He first established (11) for the case of rational number field.

2. Let $\mathfrak{p}$ denote a prime ideal of K and let ℓ be a rational integer ≥ 1. Consider the congruence

$$\alpha_1\lambda_1^k + \ldots + \alpha_s\lambda_s^k \equiv 0 \pmod{\mathfrak{p}^\ell}. \tag{12}$$

A set of numbers $\lambda_1,\ldots,\lambda_s$ of K satisfying (12) is called a nontrivial solution of (12) if $\lambda_1,\ldots,\lambda_s$ are integers, not all divisible by $\mathfrak{p}$. Let $\Gamma^*(k,K,\mathfrak{p}^\ell)$ be the least number such that if $s \geq \Gamma^*(k,K,\mathfrak{p}^\ell)$ and $\alpha_1,\ldots,\alpha_s$ are any given integers in K, the congruence (12) has a nontrivial solution. Further, let

$$\Gamma^*(k,K) = \max_{\mathfrak{p},\ell} \Gamma^*(k,K,\mathfrak{p}^\ell).$$

To prove (3) we need an upper estimation for $\Gamma^*(k,K)$. We proved

$$\Gamma^*(k,K) \leq \begin{cases} ckn\log k, & \text{if } 2\nmid k, \\ (2k)^{n+1}, & \text{if } k \geq 1, \end{cases}$$

where c is an absolute constant (see [10]). This gives an improvement of a result $\Gamma^*(k,K) \leq 4k^{2n+3}+1$ due to Peck [1].

References

[1] L.G. Peck, Diophantine equations in algebraic number fields, Amer. J. Math. 71, 1949, 387-402.

[2] W.M. Schmidt, Small zeros of additive forms in many variables, Trans. Amer. Math. Soc. 248, 1, 1979, 121-133.

[3] W.M. Schmidt, Small zeros of additive forms in many variables II, Acta Math. 143, 1979, 219-232.

[4] W.M. Schmidt, Diophantine inequalities for forms of odd degrees, Adv. in Math. 38, 1980, 128-151.

[5] C.L. Siegel, Generalization of Waring's problem to algebraic number fields, Amer. J. Math. 66, 1944, 122-136.

[6] C.L. Siegel, Sums of m-th powers of algebraic integers, Ann. of Math. 46, 1945, 313-339.

[7] Wang Yuan, Bounds for solutions of additive equations in an algebraic number field I, Acta Arith. 48, 2 (to appear).

[8] Wang Yuan, Bounds for solutions of additive equations in an algebraic number field II, Acta Arith. 48, 4 (to appear).

[9] Wang Yuan, Diophantine inequalities for forms in an algebraic number field, (to appear).

[10] Wang Yuan, On homogeneous additive congruences, (to appear).

Institute of Mathematics
Academia Sinica
Beijing, China

Contemporary Mathematics
Volume **77**, 1988

SOME RESULTS OF MODULAR FORMS

Pei Dingyi and Feng Xuning

§1. Eisenstein series of weight 3/2.

Let

$$\theta(z) = \sum_{n=-\infty}^{+\infty} e(n^2 z),$$

where z is a variable on the (complex) upper half plane H and $e(z) = e^{2\pi iz}$. It is known that

$$\theta(r(z)) = (\tfrac{c}{d})\varepsilon_d^{-1}(cz+d)^{1/2}\theta(z)$$

for all $r = \begin{pmatrix} a & b \\ c & d \end{pmatrix} \in \Gamma_0(4)$, where $r(z) = (az+b)(cz+d)^{-1}$,

$$\varepsilon_d = \begin{cases} 1 & d \equiv 1 \pmod 4 \\ i & d \equiv 3 \pmod 4), \end{cases}$$

and

$$\Gamma_0(N) = \left\{\begin{pmatrix} a & b \\ c & d \end{pmatrix} \in SL_2(\mathbb{Z}) \,\middle|\, N|c\right\}$$

for any positive integer N. The symbol $(\frac{c}{d})$ is the quadratic residue, which is slightly different from the traditional one.

We put

$$j(r,z) = (\tfrac{c}{d})\varepsilon_d^{-1}(cz+d)^{1/2} \quad \text{for} \quad r = \begin{pmatrix} a & b \\ c & d \end{pmatrix} \in \Gamma_0(4).$$

Let κ denote an odd positive integer, N a positive integer divisible by 4, and ω a Dirichlet even character module N. A holomorphic function $f(z)$ on H is called a modular form of weight $\kappa/2$ and character ω with respect to $\Gamma_0(N)$ if

(i) $f(r(z)) = \omega(d)j(r,z)^{\kappa}f(z)$ for all $r = \begin{pmatrix} a & b \\ c & d \end{pmatrix} \in \Gamma_0(N)$.

0271-4132/88 $1.00 + $.25 per page

(ii) $f(z)$ is holomorphic at the cusps of $SL_2(\mathbb{Z})$.

This definition was first introduced by G. Shimura, [7].

Let $M(N,\kappa/2,\omega)$ denote the complex vector space of all such $f(z)$. The cuspform subspace in $M(N,\kappa/2,\omega)$ and its orthogonal complement under the Petersson inner product are denoted by $S(N,\kappa/2,\omega)$ and $\mathcal{E}(N,\kappa/2,\omega)$ respectively. It has been proved that if $\kappa \geq 5$ or $\kappa = 1$, the subspace $\mathcal{E}(N,\kappa/2,\omega)$ is generated by Eisenstein series. Pei [5] proves that this is also true for the case $\kappa = 3$. Let us briefly describe this result.

Define a group extension G of $GL_2^+(\mathbb{R})$ which consists of all the pairs $\{r,\phi(z)\}$, where $r = \begin{pmatrix} a & b \\ c & d \end{pmatrix} \in GL_2^+(\mathbb{R})$ and $\phi^2(z) = \alpha \cdot \det(r)^{-1/2}(cz+d)$ with $|\alpha| = 1$. The multiplication law in G is given by

$$\{r_2,\phi_1(z)\}\cdot\{r_2,\phi_2(z)\} = \{r_1r_2,\phi_1(r_2(z))\phi_2(z)\}.$$

If $\xi = \{r,\phi(z)\} \in G$, $f \in M(N,\kappa/2,\omega)$, the action of ξ on f is defined by

$$(f|\xi)(z) = \phi(z)^{-\kappa}f(r(z)).$$

Let us now introduce the functions

$$E(s,\omega,N)(z) = y^{s/2} \sum_{r\in\Gamma_\infty\backslash\Gamma_0(N)} \omega(d_r)j(r,z)^{-3}|j(r,z)|^{-2s}$$

and

$$E'(s,\omega,N)(z) = E(s,\omega,N)(-1/Nz)z^{-3/2},$$

where

$$\Gamma_\infty = \left\{\pm\begin{pmatrix} 1 & m \\ 0 & 1 \end{pmatrix} \Big| m \in \mathbb{Z}\right\},$$

d_r is the lower right entry of r, $z \in H$ and $s \in \mathbb{C}$. If $\mathrm{Re}\, s > 1/2$, we know that $E(s,\omega,N)$ is convergent. One can show that $E(s,\omega,N)(z)$ can be continued as a meromorphic function in s to the whole complex plane, and it is holomorphic in s at $s = 0$. Moreover, if $\omega^2 \neq \mathrm{id}$, $E(0,\omega,N)(z)$ is also holomorphic in z.

Now, if $\omega^2 \neq \mathrm{id}$, we define an Eisenstein series $E(\omega,N)(z)$ by

$$E(\omega,N)(z) = E(0,\bar{\omega},N)(z).$$

If D is an odd square-free integer, we define Eisenstein series $f(\mathrm{id},4D)(z)$ and $f(\mathrm{id},8D)(z)$ by

$$f(\mathrm{id},4D)(z) = E(0,\mathrm{id},4D)(z) - (1-i)(4D)^{-1}E'(0,(\tfrac{D}{\cdot}),4D)(z),$$

$$f(\mathrm{id},8D)(z) = E(0,\mathrm{id},8D)(z) - (1-i)(8D)^{-1}E'(0,(\tfrac{2D}{\cdot}),8D)(z).$$

One can show that $E(\omega,N) \in \mathcal{E}(N,3/2,\omega)$, $f(\mathrm{id},4D) \in \mathcal{E}(4D,3/2,\mathrm{id})$ and $f(\mathrm{id},8D) \in \mathcal{E}(8D,3/2,\mathrm{id})$. Our main result is the following:

Theorem 1. For any level N and any even character ω, the space $\mathcal{E}(N,3/2,\omega)$ is generated by the Eisenstein series $E(\psi,m)$ with $\psi^2 \neq \mathrm{id}$, $f(\mathrm{id},4D)$, $f(\mathrm{id},8D)$ for certain ψ, m, D, and their transforms under certain elements $\{r,\phi(z)\}$ $(r \in GL_2(\mathbb{Z}))$ of G.

In the proof of the theorem, a basis of $\mathcal{E}(N,3/2,\omega)$ is constructed. Suppose that $N = 2^{e(z)}N'$ with odd N'. Let $r(\omega)$ be the conductor of ω. Then we have, if $e(2) = 2$,

$$\dim \mathcal{E}(N,3/2,\omega^{-1}) = 2 \sum_{\substack{c \mid N' \\ (c,N'/c) \mid N/r(\omega^{-1})}} \phi((c,N'/c)) - \dim \mathcal{E}(N,1/2,\omega^{-1}),$$

if $e(2) = 3$,

$$\dim \mathcal{E}(N,3/2,\omega^{-1}) = 3 \sum_{\substack{c \mid N' \\ (c,N'/c) \mid N/r(\omega^{-1})}} \phi((c,N'/c)) - \dim \mathcal{E}(N,1/2,\omega^{-1}),$$

and if $e(2) \geq 4$,

$$\dim \mathcal{E}(N,3/2,\omega^{-1}) = \sum_{\substack{c \mid N \\ (c,N/c) \mid N/r(\omega^{-1})}} \phi((c,N/c)) - \dim \mathcal{E}(N,1/2,\omega^{-1}),$$

where $\phi(n)$ is the Euler function.

The result of J.P. Serre and H.M. Stark about modular forms of weight 1/2 makes it possible to calculate $\dim \mathcal{E}(N,3/2,\omega)$. By that result, $\dim \mathcal{E}(N,1/2,\omega)$ is equal to the number of pairs (ψ,t), where t is an integer ≥ 1, and ψ is a totally even primitive character with conductor $r(\psi)$, such that

(i) $4r^2(\psi)t$ divides N,

(ii) $\omega(n) = \psi(n)(\frac{t}{n})$ for all n prime to N.

To construct a basis of $\dim \mathcal{E}(N,3/2,\omega)$, first of all, we find a set of forms $(f_1,f_2,\ldots,f_n)$ in this space with $n = \dim \mathcal{E}(N,3/2,\omega)$. These forms are generated by $E(\psi,m)$, $f(\mathrm{id},4D)$ and $f(\mathrm{id},8D)$ under the transformations of G. In order to prove that they are linearly independent, we use their values at cusps of $\Gamma_0(N)$. If $f(z) \in \mathcal{E}(N,3/2,\omega)$, $s = d/c$ ($d,c \in \mathbb{Z}$, $c > 0$, $(c,d) = 1$), the value $V(f,s)$ of $f(z)$ at s is defined by

$$V(f,s) = \lim_{\tau\to 0}(-c\tau)^{3/2}f(\tau+d/c),$$

and $V(f,i\infty) = \lim_{z\to i\infty} f(z)$. Suppose that $(s_1,\ldots,s_t)$ is a full set of representatives of $\Gamma_0(N()$-equivalent classes of cusps. Here we have $n < t$. We can show that the matrix $(V(f_i,s_j))$ $(1 \le i \le n,\ 1 \le j \le t)$ is of full rank, therefore $f_1,\ldots,f_n$ are linearly independent. They form a basis of $\mathcal{E}(N,3/2,\omega)$. Thus the theorem can be proved.

Now suppose that $N = 4D$ with an odd square-free integer D. This case is more interesting. Let us give the basis of $\mathcal{E}(4D,3/2,\mathrm{id})$, which is found in the proof.

For any positive integer n and prime p, let $h(p,n)$ be the exponent such that $p^{h(p,n)}\| n$. Put

$$A(2,n) = \begin{cases} 4^{-1}(1-i)(1-3\cdot 2^{-(1+h(2,n))/2}), & 2\nmid h(2,n) \\ 4^{-1}(1-i)(1-3\cdot 2^{-(1+h(2,n))/2}), & 2\mid h(2,n),\ n/2^{h(z,n)} \equiv 1 \pmod 4 \\ 4^{-1}(1-i)(1-2^{-h(2,n)/2}), & 2\mid h(2,n),\ n/2^{h(2,n)} \equiv 3 \pmod 8 \\ 4^{-1}(1-i), & 2\mid h(2,n),\ n/2^{h(2,n)} \equiv 7 \pmod 8 \end{cases}$$

and, for $p \ne 2$,

$$A(p,n) = \begin{cases} p^{-1} - (1+p)p^{-(3+h(p,n))/2}, & 2\nmid h(p,n) \\ p^{-1} - 2p^{-1-h(p,n)/2}, & 2\mid h(p,n),\ \left(\frac{-n/p^{h(p,n)}}{p}\right) = -1 \\ p^{-1}, & 2\mid h(p,n),\ \left(\frac{-n/p^{h(p,n)}}{p}\right) = 1. \end{cases}$$

Define

$$\lambda(n,4D) = L_{4D}(2,\mathrm{id})^{-1}L_{4D}(1,(\tfrac{-n}{\cdot}))\sum\mu(a)(\tfrac{-n}{a})a^{-1}b^{-1},$$

where

$$L_N(s,\omega) = \sum_{(n,N)=1} \omega(n)n^{-s},$$

and the last sum is extended over all positive integers a and b prime to $4D$ such that $(ab)^2 \mid n$, μ denotes the Möbius function.

Suppose that D has ν different prime factors, then

$$\dim \mathcal{E}(4D,3/2,id) = 2^{\nu+1} - 1.$$

The following forms form a basis of $\dim \mathcal{E}(4D,3/2,id)$:

$$g(id,4D,4D) = 1 - 4\pi(1+i)\sum_{n=1}^{\infty}\lambda(n,4D)(A(2,n)-4^{-1}(1-i))$$

$$\times \prod_{p|D}(A(p,n)-p^{-1})n^{1/2}e(nz),$$

$$g(id,4m,4D) = -4\pi(1+i)\sum_{n=1}^{\infty}\lambda(n,4D)(A(2,n)-4^{-1}(1-i))$$

$$\times \prod_{p|m}(A(p,n)-p^{-1})n^{1/2}e(nz),$$

for $m|D$, $m \neq D$,

$$g(id,m,4D) = 2\pi\sum_{n=1}^{\infty}\lambda(n,4D)\prod_{p|D}(A(p,n)-p^{-1})n^{1/2}e(nz),$$

for $m|D$, $m \neq 1$.

If $f(z) = \sum_{n=0}^{\infty} a(n)e(nz)$ is a form $M(N,3/2,\omega)$ and $f|T(p^2) = \sum_{n=0}^{\infty} b(n)e(nz)$, where $T(p^2)$ is the Heche operator corresponding to the prime p, then we have

$$b(n) = a(p^2n) + \omega(p)(\frac{-n}{p})a(n) + \omega(p^2)pa(n/p^2).$$

We understand that $a(n/p^2) = 0$ if n is not divisible by p^2. In particular, if $p|N$, then $b(n) = a(p^2n)$.

We can show that $\{g(id,4m,4D)|m|D\} \cup \{g(id,m,4D)|m|D,\ m \neq 1\}$ are all the common eigenfunctions of the Heche operators $T(p^2)$. For example, we have

$$g(id,4m,4D)|T(p^2) = g(id,4m,4D), \quad (p|2m)$$

$$g(id,4m,4D)|T(p^2) = pg(id,4m,4D), \quad (p|Dm^{-1})$$

$$g(id,4m,4D)|T(p^2) = (1+p)g(id,4m,4D), \quad (p\nmid 2D).$$

Put $g(id,4m,4D) = \sum_{n=0}^{\infty} a(n)e(nz)$, thus

$$\sum_{n=1}^{\infty} a(tn^2)n^{-s} = a(t) \prod_{p|2m} (1-p^{-s})^{-1} \prod_{p|Dm^{-1}} (1-p^{1-s})^{-1}$$

$$\times \prod_{p\nmid 2D} (1-(\frac{-t}{p})p^{-s})(1-(1+p)p^{-s}+p^{1-2s})^{-1}$$

$$= a(t)L_{Dm^{-1}}(s,id)L_{2m}(s-1,id)L_{2Dt}(s,(\frac{-t}{\cdot}))^{-1}$$

is the Euler product, where t is a positive integer.

Now we give an application of the above result [6]. Let $N(a,b,c,n)$ denote the number of solutions $(x,y,z) \in \mathbb{Z}^3$ of the diophantine equation

$$ax^2 + by^2 + cz^2 = n,$$

where a,b,c,n are positive integers and $(a,b,c) = 1$.

Set

$$f(a,b,c,z) = \theta(az)\theta(bz)\theta(cz),$$

then we have

$$f(a,b,c,z) = 1 + \sum_{n=1}^{\infty} N(a,b,c,n)e(nz).$$

It is easy to verify that $f(a,b,c,z) \in M(4[a,b,c],3/2,(\frac{abc}{\cdot}))$, where $[a,b,c]$ is the least common multiple of a,b,c. In certain special cases of a,b,c, we can have $M(4[a,b,c],3/2,(\frac{abc}{\cdot})) = \mathcal{E}(4[a,b,c],3/2,(\frac{abc}{\cdot}))$, therefore a basis of the latter is also a basis of the former. If we can express $f(a,b,c,z)$ as a linear combination of the basis, then we can find the expression for $N(a,b,c,n)$.

Let us take the case $a = 1$, $b = c = 3$ as an example. In this case $f(1,3,3,z)$ belongs to $M(12,3/2,id)$ which has a basis consisting of $g(id,12,12)$, $g(id,4,12)$ and $g(id,3,12)$. Their values at the cusps $1/3$, $1/4$ and $1/12$ are derived as follows

	1/3	1/4	1/12
g(id, 12, 12)	$\frac{i-1}{4}$	-1/3	1
g(id, 4, 12)	0	1	0
g(id, 3, 12)	$\frac{i-1}{4}$	0	0
f(1, 3, 3, z)	$\frac{1-i}{4}$	-1/3	1.

Hence

$$f(1,3,3,z) = g(id,12,12) - 2g(id,3,12).$$

Therefore we obtain

$$N(1,3,3,n) = 2\pi n^{1/2}(1/3 - A(3,n))(2(1+i)A(2,n)+1)\lambda(n,12).$$

§2. An analog of $\eta(z)$ in the Hilbert modular case.

It is well known that

$$\eta(z) = e(z/24)\prod_{n=1}^{\infty}(1-e(nz)) = \sum_{m=1}^{3}\left(\frac{3}{m}\right)e(m^2z/24)$$

is a modular form of weight 1/2.

Let F be a totally real algebraic number field, $\mathfrak{O}$ the maximal order. We construct a Hilbert modular form of weight 1/2, which is analogous to $\eta(z)$, with respect to the full Hilbert modular group $SL_2(\mathfrak{O})$.

The group $SL_2(\mathfrak{O})$ is generated by $\begin{bmatrix} 0 & 1 \\ -1 & 0 \end{bmatrix}$ and $\begin{bmatrix} 1 & b \\ 0 & 1 \end{bmatrix}$ with $b \in \mathfrak{O}$. So, for a function f, it is enough to consider the action of $\begin{bmatrix} 0 & 1 \\ -1 & 0 \end{bmatrix}$ and $\begin{bmatrix} 1 & b \\ 0 & 1 \end{bmatrix}$ on f.

Let $\mathfrak{J}$ be an integral ideal of F, d the difference of $F/\mathbb{Q}$, w a primitive ideal character with conductor $\mathfrak{J}\mathfrak{p}$, where $\mathfrak{p}$ is the product of some archimedean prime of F. Suppose for $b \in \mathfrak{O}$

$$w(b\mathfrak{O}) = \operatorname{sgn}(b)^r = \prod_{i=1}^{n}\operatorname{sgn}(b^{(i)})^{r_i} \quad \text{if} \quad b \equiv 1 \pmod{\mathfrak{J}},$$

where

$$r = (r_1,\dots,r_n) \in \mathbb{Z}^n, \quad r_i = 0 \text{ or } 1.$$

Denote $|r| = \sum_{i=1}^{n} r_i$. For $b \in F^\times$ set

$$w_0(b) = \begin{cases} w(b,\mathfrak{O})\operatorname{sgn}(b)^r & \text{if} \quad (b,\mathfrak{J}) = 1, \\ 0 & \text{if} \quad (b,\mathfrak{J}) \neq 1. \end{cases}$$

This is a primitive character of $\mathfrak{O}/\mathfrak{J}$.

In order to consider the Gauss sum $\tau(\omega)$ of ω, we use the functional equation

$$R(s,\omega) = W(\omega)R(1-s,\omega^{-1}),$$

where

$$R(s,\omega) = N(\mathfrak{J}\mathfrak{d})^{s/2}\Gamma_\omega(s)L(s,\omega)$$

$$L(s,\omega) = \prod_{\mathfrak{p}}(1-\omega(\mathfrak{p})N(\mathfrak{p})^{-s})^{-1},$$

$\Gamma_\omega(s)$ is a certain gamma factor, and

$$W(\omega) = (-i)^{|r|}\tau(\omega)N(\mathfrak{J})^{-1/2}$$

If $\omega^2 = 1$, then $W(\omega) = 1$. We have

Lemma 1. If $\omega^2 = 1$, then $\tau(\omega) = i^{|r|}N(\mathfrak{J})^{1/2}$.

Put

$$f(u,y) = \exp[-\pi \mathrm{Tr}(yu^2)] \quad \text{for} \quad u \in \mathbb{R}^n,\ y \in \mathbb{R}^n,\ y \gg 0,$$

where

$$\mathrm{Tr}(yu^2) = \sum_{i=1}^{n} y_i u_i^2.$$

For an element $z = (z_1, z_2, \dots, z_n)$ of the ring $\mathbb{C}^n$, set

$$\mathrm{Tr}(z) = \sum z_k,\ e_F = \exp(z) = \exp[2\pi i \mathrm{Tr}(z)], \qquad N(z) = \pi z_k.$$

We also identify R^n with its dual by means of $(x,y) \longmapsto e_F(-xy)$. Moreover, denoting the isomorphisms of F into R by $a \longmapsto a^{(k)}$ $(k = 1,2,\dots,n)$, we identify F with its image of the imbedding

$$F \ni a \longmapsto (a^{(1)}, a^{(2)}, \dots, a^{(n)}) \in R^n \subseteq \mathbb{C}^n.$$

Note that $\mathrm{Tr}(a) = \mathrm{Tr}_{F/Q}(a)$ and $N(a) = N_{F/Q}(a)$ for $a \in F$. it is easy to see that the Fourier transformation of f, as a function of u, is

$$\hat{f}(u,y) = (y_1 \cdots y_n)^{-1/2} f(u, 1/y).$$

Set

$$e_F(u) = \exp(2\pi i \mathrm{Tr}(u)) \quad \text{for} \quad u \in \mathfrak{D},$$

$$e_F(z) = \exp[2\pi i \left[\sum_{k=1}^{n} z_k\right]],$$

$$N(z) = \prod_{i=1}^{n} z_i, \quad \text{for} \quad z = (z_1, \dots, z_n) \in \mathbb{C}^n.$$

By the Poisson summation formula we get

<u>Lemma 2</u>. For $u \in F$ and an ideal $\mathfrak{a}$ of $\mathfrak{D}$, we have

$$\sum_{a \in \mathfrak{a}} e_F[(a+u)^2 z/2] = \mu(\mathfrak{a})^{-1} N(-iz)^{-1/2} \sum_{b \in \tilde{\mathfrak{a}}} e_F(-b^2/2z) e_F(bu),$$

where $z \in H^n$ and

$$\tilde{\mathfrak{a}} = \{x \in F \mid Tr(xy) \in \mathbb{Z}, \ \forall \ y \in \mathfrak{a}\}.$$

From Lemmas 1 and 2 we obtain the main result.

<u>Theorem 2</u>. Let ω, ω_0 and $\mathfrak{J}$ be the same as above. Suppose

(1) $\omega^2 = 1$;

(2) $\mathfrak{J}\mathfrak{d}$ is a principal ideal generated by a totally positive number, say δ;

(3) $\mathfrak{J}$ is divisible by every prime factor of 2;

(4) $u^2 \equiv 1 \pmod{2\mathfrak{J}}$ for every $u \in \mathfrak{D}$ prime to $\mathfrak{J}$.

Then

$$f(z,\omega) = \sum_{u \in \mathfrak{D}} \omega_0(u) e_F(u^2 z/2\delta), \quad (z \in H^n)$$

is a Hilbert modular form of weight 1/2. Moreover, for every $r \in \begin{pmatrix} a & b \\ c & d \end{pmatrix} \in SL_2(\mathfrak{D})$ we have

$$f(r(z),\omega) = j^*(r,z) f(z,\omega),$$

$$j^*(r,z) = \mathcal{E}^*(r,z)(cz+d)^{1/2}$$

with a root of unity $\mathcal{E}^*(r,z)$ and

$$(cz+d)^{1/2} = \prod_{k=1}^{n} (c^{(k)} z_k + d^{(k)})^{1/2}.$$

In fact, we have

$$j^*(r,z) = \begin{cases} (-i)^{|r|} N(-iz)^{1/2}, & \text{if} \quad r = \begin{pmatrix} 0 & 1 \\ -1 & 0 \end{pmatrix} \\ e_F(b/2\delta), & \text{if} \quad r = \begin{pmatrix} 1 & b \\ 0 & 1 \end{pmatrix} \quad b \in \mathfrak{D}. \end{cases}$$

In particular, we apply Theorem 2 to the real quadratic field $F = \mathbb{Q}(\sqrt{d})$, with a square-free positive integer d. Observe that the different of the field $\mathbb{Q}(\sqrt{d})$ is $(\sqrt{d})$. We obtain

Theorem 3. Let $F = \mathbb{Q}(\sqrt{d})$ with a positive square-free integer d. Suppose F has a unit λ such that $N(\lambda) = -1$. We can take λ so that $\lambda\sqrt{d}$ is totally positive. Set, for $k = 1,2$

$$f_k(z) = \sum \omega_{k0}(u) e_F[u^2 z/(2c_k \lambda\sqrt{d})],$$

where $c_1 = 4$, $c_2 = 12$, and

$$\omega_{10}(u) = \begin{cases} (-1)^{[N(u)-1]/2}, & \text{for } (u,2) = 1, \\ 0, & \text{for } (u,2) \neq 1. \end{cases}$$

$$\omega_{20}(u) = \begin{cases} \left(\frac{3}{N(u)}\right), & \text{for } (u,6) = 1, \\ 0, & \text{for } (u,6) \neq 1. \end{cases}$$

If $d \equiv 1 \pmod 8$, then $f_1(z)$ is a Hilbert modular form of the type described in Theorem 2. If further $d \equiv 1 \pmod{24}$ the same is true for $f_2(z)$.

Remark. We have obtained $f_1(z)$ for $d = 17,41,73,89,97$ and $f_2(z)$ for $d = 73,97$, etc.

§3. A system of matrix diophantine equations.

In [2], H.D. Kloosterman has proposed to find out the numbers of integral solutions $(x_1, x_2, \ldots, x_s)$ for the following system of Diophantine equations:

$$\begin{cases} x_1^2 + x_2^2 + \ldots + x_s^2 = n, \\ x_1 + x_2 + \ldots + x_s = m. \end{cases}$$

Using the theory of modular forms, he developed a formula with an error term, and an exact formula when $s = 3,5,7$, for the numbers of solutions. Note that the so-called Kloosterman sum was introduced in [2].

In [3] Lu Hongwen generalized the above problem as follows:

Let S be an m-row positive-definite integral matrix; N an $n\times m$ integral matrix; T an n-row positive-definite integral matrix, Q an n-row integral matrix. Moreover, the numbers of the $m\times n$ integral matrix G satisfying the following system of matrix Diophantine equations

$$G'SG = T, \quad NSG = Q$$

are assumed to be $\alpha(S,N;T,Q)$, where G' is the transposition of G. We attempt to gain an exact or asymptotic formula for $\alpha(S,N;T,Q)$. Also, by using the theory of Siegel modular function, Lu proved the following theorem.

Theorem 4. Let the rank of N be n, NS a prime matrix, i.e., it may be supplemented to be an m-row integral unimodular matrix. Again, we suppose $T-(S[N'])^{-1}[Q] > 0$, *and* $\min(T-(S[N'])^{-1}[Q]) \geq a|T-(S[N'])^{-1}[Q]|^{1/2}$. Then, if $m > 4n+2$ and $|T-(S[N'])^{-1}[Q]| > \left(\frac{\sqrt{3}}{2}\right)^{-n}$, we have

$$\alpha(S,N;T,Q) = \rho(T,Q) + O\left(|T-(S[N'])^{-1}[Q]|^{\frac{m-n}{2}-\frac{n+1}{2}-\frac{m-4n-2}{4n}-\frac{n-1}{2}}\right),$$

$$\rho(T,Q) = \frac{\pi^{\frac{mn}{2}-\frac{n(n+1)}{4}}}{\prod_{\nu=0}^{n-1}\Gamma\left(\frac{m-n}{2}-\frac{\nu}{2}\right)}|S|^{-n/2}\cdot|S[N']|^{-n/2}\cdot|T-(S[N'])^{-1}[Q]|^{\frac{m-n}{2}-\frac{n+1}{2}}\mathfrak{S}(T,Q),$$

$$\mathfrak{S}(T,Q) = \sum_{\{R,R_*\}_n(\mathrm{mod}\ 1)}(|\bar{R}_*|\cdot|\bar{R}|)^{-m}\sum_{G_0(\mathrm{mod}\ \bar{R}_*'\bar{R})}\eta(-S[G_0]R-NSG_0R_*)\cdot\eta(TR+QR_*),$$

where $|M|$ denotes the determinant of M; $\eta(M) = \exp(2\pi i\ \mathrm{trace}(M))$; $S[G] = G'SG$; the positive constant a and O-constant depend only on S,N,m and n; $\min T = \min_{x'x=1} T[x]$ (where x runs through the $n\times 1$ nonzero real vectors); R indicates an n-row rational symmetric matrix; R_* an n-row rational matrix; $\bar{R}$ and $\bar{R}_*$ the denominators of R and R_*, respectively, i.e., both $\bar{R}R$ and $\bar{R}_*R_*$ are integral matrices as well as $\bar{R}$ and $\bar{R}_*$ which are prime to $\bar{R}R$ and $\bar{R}_*R_*$ from left, respectively; and $\{R,R_*\}_n$ the set of pairs of the n-row rational matrices R and R_* satisfying $|\bar{R}_*| \neq 0$, $|\bar{R}| \neq 0$, and $\bar{R}_*'\bar{R} = \bar{R}'\bar{R}_*$. By mod 1, we mean that the elements of the matrices are regarded as mod 1; $G_0(\mathrm{mod}\ \bar{R}_*'\bar{R})$ implies that G_0 runs through all the integral matrices, and the two matrices G_0 and G_{01} of them are considered to be the same if and only if $G_0 - G_{01} = A\bar{R}_*'\bar{R}$, where A is an integral matrix.

The formula, when $n = 1$, becomes simpler while the $\rho(T,Q)$ of $m = 3,5,7$ identify the exact ones obtained by H.D. Kloosterman.

Note that the higher dimensional Kloosterman sum was also first introduced by Lu in [3].

§4. The theory of Hamiltonian modular function.

Following C.L. Siegel, for the ordinary quaternion algebra, Lu studied the theory of Hamiltonian modular function in [4]. The main results are as follows.

1. The reductive theory of the positive definite Hamiltonian matrix was constructed.

2. Lu constructed the Hamiltonian symplectic transformation group for the Hamiltonian upper half space, Hamiltonian symplectic metric and the fundamental domain for Hamiltonian modular group. It was proved that the Hamiltonian modular group is the first kind of discontinued subgroup of Hamiltonian symplectic group.

3. The Poincaré series and Eisenstein series were constructed and two theorems on convergence of both series were proved.

4. An analytic identity, which is equivalent to the "Mass formula," for Hamiltonian forms was given.

References

[1] Feng, Xuning, An analog of $\eta(z)$ in the Hilbert modular case, J. of Number Theory vol. 17, No. 1, (1983), 116-126.

[2] Kloosterman, H.D., Simultane Darstellung Zweier Zahlen als einer summe von ganzen Zahlen und deren Quadratsumme, Math. Ann. 113 (1941-1943), 319-364.

[3] Lu, Hongwen, A system of matrix Diophantine equations, Scientia Sinica 22 (1979), 1347-1361.

[4] Lu, Hongwen, The theory of Hamiltonian modular function (in Chinese), J. of Univ. Sci. & Tech. of China, v. 10, no. 4 (1980), 31-51.

[5] Pei, Dingyi, Eisenstein series of weight 3/2 I,II, Trans. of Amer. Math. Soc. 274 (1982), 573-606 and 283 (1984), 589-603.

[6] Pei, Dingyi, The number of solutions for the Diophantine equation $ax^2 + by^2 + cz^2 = n$, Kexue Tongbao, 28 (1983), 1163-1167.

[7] Shimura, G., On modular forms of half integral weight, Ann. of Math. (2) 97 (1973), 440-481.

Institute of Applied Mathematics
Academia Sinica, Beijing

Contemporary Mathematics
Volume 77, 1988

SOME RESULTS IN THE APPLICATION OF THE NUMBER THEORY TO DIGITAL SIGNAL PROCESSING AND PUBLIC-KEY SYSTEMS

Sun Qi

1. Number-theoretic transform (NTT).

In 1972-1975, Rader, Agarwal and Burrus introduced a so-called Number-theoretic Transform. This transform has the cyclic convolution property and therefore can be used for evaluating digital filters in the same way as Discrete Fourier Transforms (DFT).

If we have a sequence $x_0, x_1, \ldots, x_{N-1}$ of length N, and $\mathbb{Z}_M$ a residue classes ring of integers mod M, such that a transform

(1) $$X_k = \sum_{n=0}^{N-1} x_n \alpha^{nk}, \quad k = 0, 1, \ldots, N-1, \ \alpha \in \mathbb{Z}_M,$$

which is reversible and having cyclic convolution property, i.e., $Y_k = X_k \cdot H_k$, where $H_k = \sum_{n=0}^{N-1} h_n \alpha^{nk}$, $Y_k = \sum_{n=0}^{N-1} y_n \alpha^{nk}$, $y_k = \sum_{t=0}^{N-1} y_t h_{\langle k-t\rangle_N}$ $(k = 0, 1, \ldots, N-1)$, then (1) is called a DFT of length N over $\mathbb{Z}_M$ or a NTT of length N. For $M = q_1^{\ell_1} \cdots q_r^{\ell_r}$, where $q_1, \ldots, q_r$ are distinct primes, (Shuh Ting [1], [2] proved that 1) DFT of length N over $\mathbb{Z}_M$ exists if and only if $\alpha^N \equiv 1 \pmod M$ and that α is a primitive N-th root of unity in $\mathbb{Z}_{q_i}$ $(i = , \ldots, r)$. 2) The number of DFT of length N over $\mathbb{Z}_M$ is $\varphi^r(N)$, where $\varphi(N)$ denotes the Euler function. They [3] also have eliminated an unnecessary condition from the usual definition of a NTT.

2. m-dimensional DFT over $\mathbb{Z}_M$.

Let $X_{i_1, i_2} \in \mathbb{Z}_M$ $(i_1 = 0, 1, \ldots, N_1-1;\ i_2 = 1, \ldots, N_2-1)$ and $\alpha, \beta \in \mathbb{Z}_M$. If a two-dimensional transform

(2)
$$X_{k_1, k_2} = \sum_{n_1=0}^{N_1-1} \sum_{n_2=0}^{N_2-1} X_{n_1, n_2} \alpha^{n_1 k_1} \beta^{n_2 k_2} \quad (k_1 = 0, 1, \ldots, N_1-1;\ k_2 = 0, \ldots, N_2-1)$$

has an inverse transform and has cyclic convolution property, then (2) is called a two-dimensional DFT over $\mathbb{Z}_M$ or two-dimensional NTT, where first dimension is of length N_1 and the second dimension is of length N_2.

For a two-dimensional transform, Agarwal and Burrus obtained some sufficient conditions for existence. In 1976, Shuh Ting [4] proved that if $M = p_1^{\alpha_1}\cdots p_r^{\alpha_r}$, where $p_1,\dots,p_r$ are distinct primes, then (2) is a two-dimension DFT over $\mathbb{Z}_M$, where the first dimension of length N_1 and the second dimension of length N_2, if and only if 1) $\alpha^{N_1} \equiv 1 \pmod M$, $\beta^{N_2} \equiv 1 \pmod M$, α is a primitive N_1-th root of unity in Z_{p_i} $(i = 1,\dots,r)$, and β is a primitive N_2-th root of unity in Z_{p_i} $(i = 1,\dots,r)$, or 2) $[N_1,N_2] | O(M)$, where $[N_1,N_2]$ denotes the least common multiple of N_1 and N_2, and $O(M) = (p_1-1,\dots,p_r-1)$. In 1978, Sun Qi [1], Jsen Tehsuen [1] and Shen Chongqi [1] eliminated an unnecessary condition from the usual definition of two-dimensional DFT over $\mathbb{Z}_M$. For m-dimensional ones over $\mathbb{Z}_M$, $M \geq 3$, they [2] also proved some similar results.

3. The DFT over residue classes ring of integers in quadratic field $Q(\sqrt{m})$.

A similar transform is defined over a ring $I_{M^2}(Q)$, where

$$I_{M^2}(\theta) = \{a + b\theta,\ a,b \in \mathbb{Z}_M\},\ 2\nmid M,$$

$$\theta = \begin{cases} \sqrt{m}, & m \equiv 2,3 \pmod 4 \\ \dfrac{\sqrt{m}+1}{2}, & m \equiv 1 \pmod 4), \end{cases}$$

and m is a square-free positive integer. In 1975-1976, Reed and Truong did some works for $m = -1$ and $M = q_1\cdots q_r$; $q_1,\dots,q_r$ are distinct primes with $\left(\frac{-1}{q_j}\right) = -1$ $(j = 1,\dots,r)$. For $M = q_1^{\ell_1}\cdots q_r^{\ell_r}$, in 1978, Sun Qi, [3], Jsen Tehsuen [3] and Shen Chongqi [3] proved that 1) let $\left(\frac{m}{q_i}\right) = -1$ $(i = 1,\dots,r)$, a DFT of length N over $I_{M^2}(\theta)$ exists, if and only if there are $\alpha_1,\dots,\alpha_r$ each of which is an N-th root of unity in $I_{q_j^{2\ell_j}}(\theta)$ and a primitive N-th root of unity in $I_{q_j^2}(\theta)$ $(j = 1,\dots,r)$ or $N | (q_1^2 - 1,\dots,q_r^2 - 1)$, 2) the number of DFT of length N over $I_{M^2}(\theta)$ is $\varphi^r(N)$.

4. The DFT over residue classes ring of integers in the cyclotomic field and algebraic number fields.

A Fourier-like transform is defined over a ring $I_{M^{\varphi(n)}}(\eta)$, where

$$I_{M^{\varphi(n)}}(\eta) = \left\{ \sum_{i=0}^{\varphi(n)-1} a_i \eta^i \,|\, a_i \in \mathbb{Z}_M \right\}, \quad M > 1, \ 2 \nmid M,$$

$n > 1$, the η is a primitive n-th root of unity and $M = q_1^{\ell_1} \cdots q_r^{\ell_r}$.

In 1978, Sun Qi [4], Jsen Tehsuen [4] and Shen Chongqi [4] proved the following two results: 1) Let $\varphi_n(x)$ be irreducible mod q_i $(i = 1, \ldots, r)$, where $\varphi_n(x)$ is the cyclotomic polynomial of index n, a DFT of length N over $I_{M^{\varphi(n)}}(\eta)$ exists, if and only if there are $\alpha_1, \ldots, \alpha_r$ each of which is an N-th root of unity in $I_{q_i^{\varphi(n)\ell_i}}(\eta)$ and a primitive N-th root of unity in $GF(q_i^{\varphi(n)})$ $(i = 1, \ldots, r)$ or $N \mid (q_1^{\varphi(n)} - 1, \ldots, q_r^{\varphi(n)} - 1)$. 2) Let $\varphi_n(x)$ be irreducible mod q_i $(i = 1, \ldots, r)$, the numbers of DFT of length N over $I_{M^{\varphi(n)}}(\eta)$ are $\varphi^r(N)$.

Analogous results are proved for arbitrary algebraic number fields by Sun Qi [5].

5. The reversible transforms having cyclic convolution property (CRT) over an arbitrary commutative ring R with identity.

Jsen Tehsuen [5], Sun Qi [6] and Shen Chongqi [5] considered the CRT over an arbitrary commutative ring R with identity. They found a necessary and sufficient condition for the existence of CRT over a general ring R. When $R = Z$, they gave the numbers of all CRT and all such CRT which are not DFT.

For m-dimensional CRT, similar results are proved by Jsen [6].

In 1980, Jsen Tehsuen [7] proved that the DFT structure of length N, is the only structure of CRT of length N over R, if and only if the number of solutions of the equation $X^N - 1 = 0$ is no more than N in the R.

In 1985, Shen Chongqi [6] obtained necessary and sufficient conditions for the existence of m-dimensional $(m \geq 33)$ reversible transforms having cyclic convolution property over an arbitrary commutative ring R with identity and their constructing principles.

6. A prime factor fast Fourier transform algorithm.

Recently, several new ideas have emerged which lead to new algorithms for the DFT. One key idea described by Rader in 1968 was the observation that the computation of the DFT can be changed into a cyclic convolution by rearranging the data when N is an odd prime.

In 1976, Winograd, Rader, and in 1979, Sun Qi [7] proved that the DFT can also be converted to cyclic convolution, when $N = p^{\ell}$ and $N = 2p^{\ell}$ respectively, where p is an odd prime, $\ell \geq 1$.

7. A kind of trap door one-way function.

In 1976, Diffie and Hallman proposed a new method of encryption called public-key systems, which are based on the trap door one-way function. In 1977, Rivest, Shamir and Adleman gave a kind of trap door one-way function based on the difficulty of factoring large numbers.

In 1985, Sun Qi [8] found a kind of new trap door one-way function. He proved the following theorem.

Suppose that n is a positive integer, p is a prime number, $n = a_h p^h + \dots + a_0$, where integers a_j satisfy $0 \leq a_j < p$ $(j = 0, 1, \dots, h-1)$, $1 \leq a_h < p$ and we write $n = [a_h, \dots, a_0]_p$. Let an arbitrary interval $[1, M]$ be given. For each arbitrary $n \in [1, M]$ it is possible to indicate an m-th polynomial $g(x) = g_1^{\ell_1}(x) \dots g_k^{\ell_k}(x)$ over F_p such that $(g(x), a_h x^h + \dots + a_0) = 1$ and $m > h$, where $\ell_j \geq 1$, $g_j(x)$ are m_j-th irreducible over F_p, $j = 1, \dots, k$. Suppose that s is a positive integer, $(s, p^m \prod_{i=1}^{k} (1 - \frac{1}{p^{m_i}})) = 1$. If we define the function $f(n) = [b_t, \dots, b_0]_p$, then $f(n)$ is a kind of trap door one-way function, where $b_t, \dots, b_0$ satisfy

$$<(a_h x^h + \dots + a_0)^s>_{g(x)} = b_t x^t + \dots + b_0,$$

and $<H(x)>_{g(x)}$ denotes the remainder polynomial of $H(x)$ modulus $g(x)$.

8. A kind of trap door one-way function over algebraic integers.

In 1986, Sun Qi [9] gave a kind of trap door one-way function over algebraic integers. He proved the following two theorems.

1)Suppose that $\mathbb{Q}(i)$ is a complex number field, $D = \{a + bi : a, b \in \mathbb{Z}\}$, where $\mathbb{Z}$ denotes the domain of rational integers. Let $m = q_1^{n_1} \dots q_k^{n_k}$, $q_j \equiv 3 \pmod 4$ $(j = 1, \dots, k)$, where $q_1, \dots, q_k$ are distinct primes. Let $s > 0$,

$$\left(s, m^2 \prod_{j=1}^{k} \left(1 + \frac{1}{q_j^2}\right)\right) = 1,$$

$\alpha \in T = \{a+bi : 0 \le a, b < m,$ and q_j $(j = 1,\dots,k)$ are prime to at least one in a and $b\}$. If we define the function

$$f(\alpha) = a' + b'i = \beta, \quad 0 \le a', b' < m,$$

then $f(\alpha)$ is a kind of trap door one-way function, where β satisfies

$$\beta \equiv \alpha^s \pmod{[m]}.$$

2) Suppose that $K = \mathbb{Q}(\theta)$ is an algebraic number field of degree n, Δ is the discriminant of K, and $\omega_1,\dots,\omega_n$ are a basis of D, where D is the domain of algebraic integers of K. Let $m = p_1 \cdots p_k$, where $p_1,\dots,p_k$ are distinct primes. Let $s > 0$, $(s, m^n \prod_{p|[m]} (1-\frac{1}{N(p)})) = 1$, where p runs through the distinct prime ideal divisors of $[m]$, $\alpha \in T_1 = \{a_1\omega_1 + \dots + a_n\omega_n : 0 \le a_j < m,\ j = 1,\dots,n\}$. If we define the function

$$f(\alpha) = \beta = b_1\omega_1 + \dots + b_n\omega_n, \quad 0 \le b_j < m, \quad j = 1,\dots,n,$$

then $f(\alpha)$ is a kind of trap door one-way function, where β satisfies

$$\beta \equiv \alpha^s \pmod{m}.$$

References

Jsen Tehsuen

[1] See Sun Qi [1].

[2] See Sun Qi [2].

[3] See Sun Qi [3].

[4] See Sun Qi [4].

[5] Acta Sci. of Sichuan Univ., 4 (1978), 1-12 (with Sun Qi and Shen Chongqi).

[6] Ibid., 3 (1979), 19-30.

[7] Ibid., 2 (1980), 17-22.

Shen Congqi

[1] See Sun Qi [1].

[2] See Sun Qi [2].

[3] See Sun Qi [3].

[4] See Sun Qi [4].

[5] See Jsen Tehsuen [1].

[6] Acta Sci. of Sichuan Univ., 3 (1976), 7-10.

Shuh Ting

[1] Practice and Knowledge in Math., 3 (1977), 45-52.

[2] Ibid., 4 (1977), 47-56.

[3] Ibid., 2 (1978), 81.

[4] Acta Sci. of Sichuan Univ., 3 (1976), 7-10.

Sun Qi

[1] Acta Sci. of Sichuan Univ., 2-3 (1978), 1-3 (with Jsen Tehsuen and Shen Chongqi).

[2] Fast number-theoretic transforms, Science Press, 1980 (with Jsen Tehsuen and Shen Chongqi).

[3] Acta Sci. of Sichuan Univ., 1 (1978), 1-10 (with Jsen Tehsuen and Shen Chongqi).

[4] Ibid., 4 (1978), 13-19 (with Jsen Tehsuen and Shen Chongqi).

[5] Ibid., 3 (1979), 11-18.

[6] See Jsen Tehsuen [1].

[7] Ibid., 2 (1979), 65-70.

[8] Kexue Tongbao, 5 (1986), 358-359.

[9] Ibid., 16 (1986), 1149.

Institute of Mathematics
Sichuan University, Chendu

Contemporary Mathematics
Volume **77**, 1988

SOME RESULTS ON DIOPHANTINE EQUATIONS

Sun Qi

1. Frobenius Problem.

Let $a_1, \ldots, a_k$ be relatively prime positive integers. Then for every sufficiently large integer n there exist <u>nonnegative</u> integers $x_1, \ldots, x_k$ such that $n = a_1x_1 + \ldots + a_kx_k$. Let $g(a_1, \ldots, a_k)$ be the largest n which is not representable in such a form. The problem of Frobenius is to determine this quantity.

For $k = 2$, $g(a_1, a_2) = a_1a_2 - a_1 - a_2$ was given by Sylvester. But for $k \geq 3$ it is an unsolved problem. Ko Chao [1],[2], Chen Zhongmu [1], Lu Wenduan [1], [2], Wu Changjiu [1], and Yin Wenlin [1] and Li Delang [1] investigated this problem and obtained several results for $g(a_1, \ldots, a_k)$.

Recently, Chen Zhongmu [2] gave a simpler algorithm to find $g(a_1, a_2, a_3)$. Let α, β, γ be three nonnegative integers, k be the least positive integer such that $k\alpha = \ell\beta + m\gamma$ ($\ell \geq 0$, $m \geq 0$), which is called normal form of α with respect to β, γ. Suppose $(a_1, a_2) = d_3$, $(a_2, a_3) = d_1$, $(a_3, a_1) = d_2$, $a_1 = \alpha d_2 d_3$, $a_2 = \beta d_1 d_3$, $a_3 = \gamma d_1 d_2$, and the normal forms of α with respect to β, γ and of β with respect to α, γ are $k\alpha = \ell\beta + m\gamma$, $u\beta = v\alpha + w\gamma$, respectively. Chen proved that if m or w is not zero, then $g(a_1, a_2, a_3) = \max(k\alpha + w\gamma,\ u\beta + m\gamma) d_1 d_2 d_3 - a_1 - a_2 - a_3$.

2. Some exponential equations.

Erdös asked for integer solutions of the equation

$$x^x y^y = z^z \tag{1}$$

with $x > 1$, $y > 1$. In 1940, Ko Chao [3] proved that when $(x,y) = 1$, equation (1) has no solutions in positive integers $x > 1$, $y > 1$, $z > 1$ and when $(x,y) > 1$, equation (1) has infinitely many solutions

$$x = 2^{2^{n+1}(2^n-n-1)+2n}(2^n-1)^{2(2^n-1)}$$

$$y = 2^{2^{n+1}(2^n-n-1)}(2^n-1)^{2(2^n-1)+2}$$

$$z = 2^{2^{n+1}(2^n-n-1)+n+1}(2^n-1)^{2(2^n-1)+1},$$

$n > 1$. Other solutions have not been found yet. Recently, Erdös pointed out that is is possible that these should be all the solutions of equation (1).

Anderson conjectured that the equation $w^w x^x y^y = z^z$ has no solution with $1 < w < x < y$.

In 1964, Ko Chao]4] and Sun Qi [1] proved the equation

$$\prod_{i=1}^{k} x_i^{x_i} = z^z, \quad x_i > 1, \; k \geq 2, \; i = 1, \ldots, k,$$

has infinitely many solutions

$$x_1 = k^{k^n(k^{n+1}-2n-k)+2n}(k^n-1)^{2(k^n-1)}$$

$$x_2 = k^{k^n(k^{n+1}-2n-k)}(k^n-1)^{2(k^n-1)+2}$$

$$x_3 = \ldots = x_k = k^{k^n(k^{n+1}-2n-k)+n}(k^n-1)^{2(k^n-1)+1}$$

$$z = k^{k^n(k^{n+1}-2n-k)+n+1}(k^n-1)^{2(k^n-1)+1}$$

of which the first one is $x_1 = 3^{14}2^4$, $x_2 = 3^{12}2^6$, $x_3 = 3^{13}2^5$, $z = 3^{14}2^5$ for $k = 3$. It gives a counter-example to Anderson's Conjecture.

For $2 \nmid xy$ are there any solutions of equation (1)? This still remains unproved. For the equation $x_1^{x_1} x_2^{x_2} x_3^{x_3} = z^z$ we asked that is there any solution with $1 < x_1 < x_2 < x_3$, $2 \nmid x_1 x_2 x_3$?

In 1957, Ko Chao [5] studied the equations $x^y y^x = z^z$, $x^y y^z = z^x$, $x^x y^z = z^y$. He proved that for $x > 1$, $y > 1$, $(x,y) = 1$, these equations have no integer solution, and for $(x,y) > 1$, each of these equations has infinitely many integer solutions. In 1958, Lu Wenduan [3] studied the equation $\prod_{i=1}^{n} x_i^{a_i} = \prod_{j=1}^{n} y_j^{b_j}$. In 1980, Yan Faxiang [1] proved that if integers $x > 1$, $y > 1$, $z > 1$ satisfy the equation $x^y y^z = z^x$, then x, y have the same prime factors.

Jesmanowicz conjectured that the diophantine equation

$$a^x + b^y = c^z$$

has no integer solution except $x = y = z = 2$, where a, b, c satisfy

$a^2 + b^2 = c^2$. Ko Chao [6]-[9] made several investigations about it in 1958-1965. Lu Wenduan [4], Chen Jingrun [1] and Sun Qi [2] also have studied this conjecture. For example, Lu proved that if $a = 4n^2-1$, $b = 4n$, $c = 4n^2+1$, then Jesmanowicz's conjecture is true.

For the Diophantine equation

$$a^x + b^y = c^z \tag{2}$$

where a, b, c are different primes, were studied by Nagell, Makowski, Hadano, and Uchiyama between 1958-1976. They gave all the solutions (x, y, z) for $\max(a, b, c) \le 17$. In 1984, Sun Qi [3] and Zhou Xiaoming [1] gave all the non-negative integral solutions of equation (2) for $\max(a, b, c) = 19$. In 1985, $\max(a, b, c) = 23$, has been solved by Yang Xiaozuo [1].

Hall asked for the integer solution of the Diophantine equation

$$p^r + 2 = q^s \tag{3}$$

where p, q are prime numbers. This includes that $5^2 + 2 = 3^3$, but we know no other case in which both $r > 1$ and $s > 1$. Sun Qi [4] and Zhou Xiaoming [2] studied the case $p + 2 = q$ in (3). Later Cao Zhenfu [1] proved that when $p + 2 = q$, then the equation (3) has no solution for $r > 1$, $s > 1$.

Selfridge asked for what a and b

$$2^a - 2^b \mid n^a - n^b, \tag{4}$$

is true for all n?

Sun Qi [5] and Zhang Mingzhi [1] proved that, for $0 \le b < a$, the equation (4) is true for all n if and only if $(a, b) = (1,0), (2,1), (3,1), (4,2), (5,3), (5,1), (6,2), (7,3), (8,4), (8,2), (9,3), (14,2), (15,3), (16,4)$.

3. Some diophantine equations connnected with arithmetical functions.

In 1963-1965, Ko Chao [10],[11] and Sun Qi [6],[7] studied some equations involving multiplicative functions. One of them is $k\varphi(n) = n - 1$, $k \ge 2$, where $\varphi(n)$ is the Euler function. In 1932, Lehmer conjectured that this equation has no positive integer solution. This is an unsolved conjecture. In 1932, Lehmer proved that if this equation has a solution, then n has at least 7 different prime factors. In 1963, Ko Chao [12] and Sun Qi [8] improved 7 to 12, and proved that $n > 2.6\times10^{17}$. Given the sequences:

$$u_1 = 1,\ u_2 = 1,\ u_{n+2} = u_{n+1} + u_n,\quad n \geq 1, \tag{5}$$

$$u_1 = 2,\ u_2 = 1,\ u_{n+2} = u_{n+1} + u_n,\quad n \geq 1, \tag{6}$$

in 1965 Ko Chao [13] and Sun Qi [9] proved that there are no squares in the sequence (5) except $u_1 = u_2 = 1$ and $u_{12} = 144$, and no squares in the sequence (6) except $u_2 = 1$, $u_4 = 4$.

In 1975, Ko Chao [14] and Sun Qi [10] proved that the equation

$$\frac{\varepsilon^{4m}+\varepsilon^{-4m}}{2} = t^2$$

has no solutions in positive integers m, t where $\varepsilon = x_0 + y_0\sqrt{D}$ is the fundamental solutions of the Pell's equation $x^2 - Dy^2 = 1$. From this result, we easily deduce that $v_0/2$ is the only square in the sequence $v_{4m}/2$ $(m = 0, 1, \ldots)$ defined by the Lucas sequence

$$v_{n+2} = 2x_0v_{n+1} - v_n,\ v_0 = 2,\ v_1 = 2x_0\quad (n = 0, 1, \ldots).$$

Recently, Sun Qi [11] studied a kind of diophantine equation connected with Lucas sequences $v_{n+2} = 2av_{n+1} - kv_n$, $v_0 = 2$, $v_1 = 2a$, $(n = 0, 1, \ldots)$, where $k = (a+b\sqrt{D})(a-b\sqrt{D})$, and a, b are integers $(b \neq 0)$

4. Some cubic equations and quartic equations.

From the well-known identity

$$(x+1)^3 + (x-1)^3 - 2x^3 = 6x,$$

Mordell suggested that perhaps most integers can be expressed as $x^3 + y^3 + 2z^3$ with integers x, y, z. In 1936, Ko Chao]15] gave the decompositions into four cubes in this form for $n \leq 100$ except the numbers 76, 99.

An interesting equation is

$$x^3 + y^3 + z^3 = n. \tag{7}$$

When $n = 3$, there are solutions given by $(x,y,z) = (1,1,1)$, $(4,4,-5)$, $(4,-5,4)$, $(-5,4,4)$. In 1984, Scarowsky and Boyarsky proved that the equation (7) has no new solutions found for $|m| \leq 50{,}000$, where $x + y + z = 3m$, $m \in Z$. In 1985, Cassels proved that any integral solution of the equation (7) satisfies $x \equiv y \equiv z \pmod 9$. Recently, Sun Qi [12] proved that if $n = 9a^3$, where a is not divisible by primes of the form $6k + 1$, then any integral solution of the equation $x^3+y^3+z^3=9a^3$ satisfies $9\mid\frac{xyz}{d^3}$, where $(x,y,z)=d$. If $n=3a^3$, $3\nmid a$, then any integral solution of the equation $x^3+y^3+z^3=3a^3$ satisfies $\frac{x}{d}\equiv\frac{y}{d}\equiv\frac{z}{d} \pmod 9$.

In 1960, Ko Chao [16], [17] and in 1962, Cassels and Sansone independently proved that the equation $x^3+y^3+z^3=xyz$ has no solutions in integers $xyz\neq 0$. From this result, we easily deduce that equation $xyz = x+y+z=1$ has no rational solutions.

Ljunggren proved that if $D>2$ is a square free integer which is not divisible by primes of the form $6n+1$, then the equation

$$x^3\pm 1 = Dy^2 \tag{8}$$

has at most one solution in positive integers x,y. In 1981, Ko Chao [18] and Sun Qi [13] proved that the only solution in integers of the equation (8) is $x=\pm 1$, $y=0$. In 1975-1982, Ko Chao [18]-[23] and Sun Qi [14]-[18] studied the equation

$$x^4-Dy^2=1,\ D>1,\ \mu(D)\neq 0. \tag{9}$$

They proved that 1) if $D\equiv 3 \pmod 8$, where $\varepsilon = x_0+y_0\sqrt{D}$ is the fundamental solution of the equation $x^2-Dy^2=1$ and if $x_0\equiv 0 \pmod 2$, then the equation (9) has no solutions in positive integers x,y. 2) If $D=2p$, where p is an odd prime number, then the equation (9) has no positive integral solutions except $p=3$, $x=7$, $y=20$. 3) If D is not divisible by primes of the form $4n+1$, then the equation (9) has no positive integral solutions. In 1979 and 1981, Ko Chao [24]-[27] and Sun Qi [19]-[22] also studied the equations $x^2-Dy^4=1$, $x^4+4=Dy^2$ and $x^2\pm 8=Dy^2$.

In 1964, Ko Chao [28], Sun Qi [23], and S.J. Chang [1], in 1981, Li Delang and in 1981, Yang Xunquian [1] studied the equation

$$\frac{4}{n}=\frac{1}{x}+\frac{1}{y}+\frac{1}{z}.$$

In 1983, Zhang Mingzhi [2] proved that if 1) $k>0$, $k\equiv 3 \pmod 8$, or 2) $k-2$ is a prime or 3) $k+2=pq$, $p\equiv 3 \pmod 8$, $q\equiv 7 \pmod 8$, p and q are primes, then the equation $x^4+kxy+y^4=z^2$ has the only trivial solutions $xy=0$.

Is $x=24$, $y=70$ the only nontrivial solution of the diophantine equation $6y=x(x+1)(2x+1)$? This was solved (affirmatively) by Watson, using elliptic functions, and by Ljunggren, using a Pell equation in a quadratic field. Mordell asked if there was an elementary proof. In 1985, Ma Degang [1] answered Mordell's question.

5. Some equations of degree > 4.

In 1842, Catalan conjectured that the only solution in integers $m > 1$, $n > 1$, $x > 1$, $y > 1$ of the equation $x^m - y^n = 1$ is $n = 3$, $m = 2$, $y = 2$, $x = 3$. In 1961, Ko Chao [29] proved that the equation $x^2 = y^n + 1$, $n > 1$, has only solutions $n = 3$, $x = \pm 3$, $y = 2$ with $xy \neq 0$. In 1960, Cassels, and in 1962 Ko Chao [30] independently proved that if the equation $x^p - y^q = 1$ has solutions $x > 1$, $y > 1$, where p and q are odd primes, then $q|x$, $p|y$. From this result, we easily deduce that no three consecutive integers are powers of integers.

In 1963, Ko Chao [31] proved that for $n > 3$ the equation

$$\sum_{j=0}^{h} (x-j)^n = \sum_{j=1}^{h} (x+j)^n$$

has no positive integer solution.

In 1962, Ko Chao [32] and Sun Qi [24] investigated the equation $x^n + \ldots + (x+h)^n = (x+h+1)^n$. In 1978, Ko Chao [33], Sun Qi [25] and Zou Zhaonan [1] proved that if n is an odd integer, then the equation has no positive integer solution except $n = 3$, $h = 2$, $x = 3$ and $n = h = x = 1$. Recently, Ko Chao [34] and Sun Qi [26] proved that if the equation has positive integer solution, then $1.1447n + 0.7417 < x + h < 1.8533n + 0.526$.

In 1979 and in 1980, Dai Zongduo [1],[2], Yu Kunrui [1],[2] and Feng Xuning [1],[2] studied the equations $x^{1/n} + y^{1/n} = z^{1/n}$ and $x^{m/n} + y^{m/n} = z^{m/n}$. They have given all positive integer solutions of these equations.

6. The equation $\frac{1}{x_1} + \ldots + \frac{1}{x_s} + \frac{1}{x_1 \ldots x_s} = 1$ and a problem of Znàm's.

For the equation

$$\sum_{i=1}^{s} \frac{1}{x_i} + \frac{1}{x_1 \ldots x_s} = 1, \quad 0 < x_1 < \ldots < x_s, \tag{10}$$

in 1964, Ko Chao [35] and Sun Qi [27] gave all solutions for $s = 5$ and $s = 6$.

Let $\Omega(s)$ be the number of positive integral solutions of the equation (10). In 1978, Sun Qi [28] proved that when $s \geq 4$, then $\Omega(s) < \Omega(s+1)$.

In 1978, Janàk and Shula gave eighteen solutions of the system of congruences

$$x_1 \dots x_{i-1} x_{i+1} \dots x_n + 1 \equiv 0 \pmod{x_i}, x_i > 1, \ i = 1, \dots, n \tag{11}$$

for $n = 7$. Let $H(n)$ be the number of solutions of the system of congruences (11). In 1983, Sun Qi [29] proved that if $n \geq 4$, then $H(n) < H(n+1)$. As a consequence we derive that: If $n \geq 7$, then $H(n) \geq n + 11$.

In 1972, Znàm asked whether for every positive integer $n > 1$ there exist integers $x_i > 1$ $(i = 1, \dots, n)$ such that x_i is a proper divisor of the numbers $x_1 \dots x_{i-1} x_{i+1} \dots x_n + 1$ for every i. In 1983, Sun Qi [30] proved that if $Z(n)$ denotes the number of solutions of the Znàm problem with $1 < x_1 < \dots < x_n$, then for $n \geq 5$, $Z(n) \geq \Omega(n) - \Omega(n-1) > 0$. Hence the problem of Znàm is completely solved. It is difficult to prove $Z(n+1) > Z(n)$, when $n \geq 5$.

In 1985, Sun Qi [31] and Cao Zhenfu [2] studied the equation

$$\sum_{i=1}^{s} \frac{1}{x_i} - \frac{1}{x_1 \dots x_s} = 1, \quad 0 < x_1 < \dots < x_s. \tag{12}$$

Let $A(s)$ be the number of solutions of the equation (12). They proved that if $t \geq 9$, then $A(t+1) \geq \Omega(t) + \Omega(t-1) + 6$.

7. Applications of Ko-Terjanian-Rotkiewicz method to diophantine equations.

Let p be an odd prime > 3, in order to prove that equation $x^2 - 1 = y^p$ has no integer solutions with $x > 1$, Ko Chao calculated the Jacobi symbol $\left(\frac{Q_p(y)}{Q_q(y)}\right)$, where $Q_n(y) = \frac{y^n+1}{y+1}$, $2 \nmid n$, and q is an odd prime with $q \neq p$ (see [29]). In 1977, Terjanian proved that equation $x^{2p} + y^{2p} = z^{2p}$, where p is an odd prime, has no integer solutions if $2p \nmid x$ and $2p \nmid y$. In order to prove this theorem Terjanian calculated the Jacobi symbol $\left(\frac{A_m(x,y)}{A_n(x,y)}\right)$ where $2 \nmid mn$, $A_i(x,y) = \frac{x^i - y^i}{x-y}$, $4 \mid x-y$. In 1983, Rotkiewicz applied ideas of Ko and Terjanian to some diophantine equations connected with Lehmer's number and calculated the Jacobi symbol $\left(\frac{P_n}{P_m}\right)$, where $2 \nmid mn$, P_n is the Lehmer number $\frac{\alpha^n - \beta^n}{\alpha - \beta}$, α and β are roots of the trinomial $z^2 - \sqrt{L}z + M$, $L > 0$ and M are rational integers.

We call the above method to be the Ko-Terjanian-Rotkiewicz method. It is elementary and useful. With this method, the following results are proved.

In 1984, Le Maohua [1], and in 1985, Zhu Weisan [1] independently proved that if the equation $x^4 - Dy^2 = 1$ has solutions in integer numbers x, y, then $x_1^2 = x_0$ or $x_0^2 = Dy_1^2$, where (x_1, y_1) is the minimal solution of $x^4 - Dy^2 = 1$ and $x_0 + y_0\sqrt{D}$ is the fundamental solution of the Pell's equation $x^2 - Dy^2 = 1$. Recently Sun Qi [32] gave a simplified proof.

Cao Zhenfu [3] proved that if p is a prime > 3, then the equation $7x^2 + 1 = y^p$ has no integer solution with $xy \neq 0$. In 1985, Sun Qi [33] proved that the equations $15x^2 + 1 = y^p$ and $23x^2 + 1 = y^p$ have no integer solutions with $xy \neq 0$.

8. The equation $\sum_{i=1}^{n} \frac{y_i}{d_i} \equiv 0 \pmod 1$.

Let $d_1, \ldots, d_n$ be fixed positive integers. It is well known that the number $I(d_1, \ldots, d_n)$ of solutions of the equation

$$\frac{y_1}{d_1} + \frac{y_2}{d_2} + \ldots + \frac{y_n}{d_n} \equiv 0 \pmod 1, \quad y_i \text{ integers}, \qquad (13)$$
$$1 \le y_i < d_i \quad (i = 1, \ldots, n)$$

play an important role in the study of diagonal equations over finite fields.

In 1948 and 1949, Hua [1],[2] and Vandiver [1],[2], Furtado, Weil at about the same time proved the following results. If N denotes the number of solutions of the equation

$$\sum_{i=1}^{n} a_i x_i^{d_i} = 0, \quad \text{where } d_i | q-1 \quad (i = 1, \ldots, n) \qquad (14)$$

over finite field F_q, then

$$|N - q^{n-1}| \le I(d_1, \ldots, d_n)(q-1)q^{(n-2)/2}. \qquad (15)$$

Hence the value of $I(d_1, \ldots, d_n)$ heavily affects the estimate of the number N of solutions of the equation (14).

For the case $d_1 = d_2 = \ldots = d_n$, it is proved that $I(d, \ldots, d) = \frac{d-1}{d}((d-1)^{n-1} + (-1)^n)$. For the more general case $d_1 | d_2 \ldots | d_n$, Sun Qi [34], Wan Daqing [1], Ma Degang [2] proved that $I(d_1, \ldots, d_n) = \prod_{j=1}^{n-1}(d_j - 1) - \prod_{j=1}^{n-2}(d_j - 1) + \ldots + (-1)^{n-1}(d_2 - 1)(d_1 - 1) + (-1)^n(d_1 - 1)$. A complicated formula for $I(d_1, \ldots, d_n)$ was obtained independently by Lide and Niederreiter, Stanly, and us [35] with different methods. The formula can be expressed as follows

$$I(d_1,\ldots,d_n) = (-1)^n + \sum_{r=1}^{n}(-1)^{n-r} \sum_{1\le i_1<\ldots<i_r\le n} \frac{d_{i1}\cdots d_{ir}}{\mathrm{lcm}[d_{i1},\ldots,d_{ir}]}.$$

For (15), it is interesting to determine when $I(d_1,\ldots,d_n) = 0$. For, if $I(d_1,\ldots,d_n) = 0$, then (14) has exactly q^{n-1} solutions. Some partial results have been obtained by Joly. In 1985, Sun Qi [36] and Wan Daqing [2] proved the following theorem: Let $n > 2$, then $I(d_1,\ldots,d_n) = 0$ if and only if one of the following conditions holds. 1) for some d_i, $(d_i, \frac{d_1\cdots d_n}{d}) = 1$, or 2) if $d_{i1},\ldots,d_{ik}$ $(1 \le i_1 < \ldots < i_k \le n)$ is the set of all even integers among $\{d_1,\ldots,d_n\}$, then $2\nmid k$, $\frac{d_{i1}}{2},\ldots,\frac{d_{ik}}{2}$ are pairwise prime, and d_{ij} is prime to any odd number in $\{d_1,\ldots,d_n\}$ $(j = 1,\ldots,k)$ if $k < n$.

References

Cao Zhenfu

[1] Nature Journal, 6 (1985), 476-477.

[2] See Sun Qi [31].

[3] Journal of the Southwest Teachers College, 2 (1985), 69-73.

Chang S.J.

[1] See Ko Chao [28].

Chen Jingrun

[1] Acta Sci. of Sichuan Univ., 2(1962), 19-25.

Chen Zhongmu

[1] Acta Sci. of Sichuan Univ., 1 (1956), 57-59.

[2] Journal of the Southwest Teachers College, 3 (1984), 1-8.

C.H. Chou

[1] See Ko Chao [33].

Dai Zongdue

[1] Acta Math. Sinica, 5 (1980), 799-793 (with Yu Kunrui and Feng Xuning).

[2] Kexue Tongbao, 10 (1979), 493-442 (with Fenb Xuning and Yu Kunrui).

Feng Xuning

[1] See Dai Zongduo [1].

[2] See Dai Zongduo [2].

Hua L.K.

[1] Proc. Nat. Acad. Sci. USA, 34 (1948), 258-263 (with Vandiver, H.S.).

[2] Ibid., 35 (1948), 94-99 (with Vandiver, H.S.).

Ko Chao

[1] Acta Sci. of Sichuan Univ., 1 (1955), 1-4.

[2] Ibid., 3 (1964), 39-50 (with Yin Wenlin and Li Delang).

[3] Journal of the Chinese Math. Soc., 2 (1940), 205-207.

[4] Acta Sci. of Sichuan Univ., 2 (1964), 5-9 (with Sun Qi).

[5] Ibid., 1 (1957), 33-40.

[6] Ibid., 1 (1958), 73-80.

[7] Ibid., 2 (1958), 81-90.

[8] Ibid., 3 (1959), 25-34.

[9] Ibid., 3 (1964), 1-12.

[10] Ibid., 1 (1963), 1-12.

[11] Ibid., 2 (1965), 1-9 (with Sun Qi).

[12] See [10].

[13] Ibid., 2 (1965), 11-18 (with Sun Qi).

[14] Ibid., 1 (1975), 57-61 (with Sun Qi).

[15] Journal of the London Math Soc., 11 (1936), 218-219.

[16] Acta Sci of Sichuan Univ., 3 (1960), 7-16.

[17] Ibid., 3 (1960), 1-6.

[18] Sci. Sin., 12 (1981), 1453-1457 (with Sun Qi).

[19] Acta Sci. of Sichuan Univ., 2 (1981), 1-6 (with Sun Qi).

[20] Acta Math. Sinica, 6 (1980), 922-926 (with Sun Qi).

[21] Chinese Annals of Math., 1 (1980), 83-89 (with Sun Qi).

[22] Kexue Tongbao, 16 (1979), 721-723 (with Sun Qi).

[23] Acta Sci. of Sichuan Univ., 3 (1980), 37-44 (with Sun Qi).

[24] Ibid., 2 (1984), 1-3 (with Sun Qi).

[25] Chinese Annals of Math., 2 (1981), 491-496 (with Sun Qi)

[26] Acta Sci. of Sichuan Univ., 4 (1979), 1-4 (with Sun Qi).

[27] Ibid., 4 (1981), 1-5 (with Sun Qi).

[28] Ibid., 3 (1964), 23-37 (with Sun Qi and S.J. Chang).

[29] Ibid., 1 (1962), 1-6.

[30] Ibid., 2 (1962), 1-6.

[31] Ibid., 1 (1963), 1-11.

[32] Ibid., 2 (1962), 9-18 (with Sun Qi).

[33] Ibid., 2-3 (1978), 19-24 (with Sun Qi and C.N. Chou).

[34] Ibid., 4 (1982), 1-4 (with Sun Qi).

[35] Ibid., 1 (1964), 13-22 (with Sun Qi).

Le Maohua

[1] Kexue Tongbao, 22 (1984), 1407.

Li Delang

[1] See Ko Chao [2].

[2] Journal of Number Theory, 4 (1981), 485-494.

Lu Wenduan

[1] Acta Sci of Sichuan Univ., 1 (1956), 49-55.

[2] Ibid., 2 (1957), 151-171 (with Wu Changjiu).

[3] Ibid., 1 (1958), 7-14.

[4] Ibid., 2 (1959), 39-41.

Ma Degang

[1] Acta Sci of Sichuan Univ., 4 (1985), 107-116.

[2] See Sun Qi [34].

Sun Qi

[1] See Ko Chao [4].

[2] See Ko Chao [9].

[3] Kexue Tongbao, 9 (1984), 1272 (with Zhou Xiaoming).

[4] See [3].

[5] Proc. AMS, 2 (1985), 218-220 (with Zhang Mingzhi).

[6] See Ko Chao [10].

[7] See Ko Chao [11].

[8] See Ko Chao [12].

[9] See Ko Chao [13].

[10] See Ko Chao [14].

[11] to appear.

[12] to appear.

[13] See Ko Chao [18].

[14] See Ko Chao [19].

[15] See Ko Chao [20].

[16] See Ko Chao [21].

[17] See Ko Chao [22].

[18] See Ko Chao [23].

[19] See Ko Chao [24].

[20] See Ko Chao [25].

[21] See Ko Chao [26].

[22] See Ko Chao [27].

[23] See Ko Chao [28].

[24] See Ko Chao [32].

[25] See Ko Chao [33].

[26] See Ko Chao [34].

[27] See Ko Chao [35].

[28] Acta Sci. of Sichuan Univ., 2-3 (1978), 15-17.

[29] Kexue Tongbao, 4 (1983), 446-447.

[30] Ibid., 11 (1983), 1564.

[31] Ibid., 5 (1985), 700 (with Cao Zhenfu).

[32] to appear.

[33] to appear.

[34] Chinese Annals of Math., 2 (1986), 232-236 (with Wan Daqing and Ma Degang).

[35] See [34].

[36] to appear in Proc. AMS (with Wan Daqing).

H.S. Vandiver

[1] See Hua, L.K. [1].

[2] See Hua, L.K. [2].

Wan Daqing

[1] See Sun Qi [34].

[2] See Sun Qi [36].

Wu Changjiu

[1] See Lu Wenduan [2].

Yan Faxiang

[1] Kexue Tongbao, 7 (1980), 543-546.

Yang Xiaozuo

[1] Acta Sci of Sichuan Univ., 4 (1985), 151-158.

Yang Xunqian

[1] Acta Sci of Sichuan Univ., 3 (1981), 101-104.

Yin Wenlin

[1] See Ko Chao [2].

Yu Kunrui

[1] See Dai Zongduo [1].

[2] See Dai Zongduo [2].

Zhang Mingzhi

[1] See Sun Qi [5].

[2] Acta Sci of Sichuan Univ., 2 (1983), 24-31.

Zhou Xiaoming

[1] See Sun Qi [3].

[2] See Sun Qi [4].

Institute of Mathematics
Sichuan University
Chengdu

Contemporary Mathematics
Volume **77**, 1988

DIOPHANTINE APPROXIMATION AND TRANSCENDENTAL NUMBER THEORY

Xu Guangshan

The study of diophantine approximation began in China in 1958. In a series of papers, Hua Lookeng and Wang Yuan proposed an algorithm of simultaneous rational approximations to a set of algebraic integers in an algebraic number field. Their method is constructive if a set of independent units of the field is known. Since a set of independent units for the cyclotomic field is known, their results have many applications in numerical analysis - in particular, in the theory of numerical integration, in multi-dimensional space. Their results will be given in another paper of Wang's [Wy].

In 1974-78, a seminar on diophantine approximation and transcendental number theory was held at the Institute of Mathematics, Academia Sinica. Since then, the study of these two topics has continued to expand in China. Now we shall give a summary of the major works of Chinese mathematicians relating to these two areas.

§1. Transference theorems.

Let θ_{ij} $(1 \le i \le m,\ 1 \le j \le n)$ be a set of mn real numbers. Let $\|x\|$ denote the distance from x to the nearest integer and $\bar{x} = \max(1, |x|)$.

Property A. For any $\varepsilon > 0$, the inequality

$$\left(\prod_{i=1}^{m} \|\theta_{i1}x_1+\dots+\theta_{in}x_n\|\right)(\bar{x}_1\cdots\bar{x}_n)^{1+\varepsilon} < 1$$

has only finitely many integral solutions $(x_1,\dots,x_n)$.

Property B. For any $\varepsilon > 0$, the inequality

$$\left(\prod_{i=1}^{n} \|\theta_{ij}y_1+\dots+\theta_{mj}y_m\|\right)(\bar{y}_1\cdots\bar{y}_m)^{1+\varepsilon} < 1$$

has only finitely many integral solutions $(y_1,\dots,y_m)$.

Schmidt and Wang [SW] proved that

Theorem 1.1. The properties A and B are equivalent.

This is a generalization of the transference theorem of Wang, Yu and Zhu [WYZ] which corresponds $m = 1$ in Theorem 1.1. The proof of Theorem 1.1 depends on a lemma of Mahler [Ma2].

A similar theorem concerning systems of linear congruences was established by Wang Yuan, Wang Lianxiang and Ren Jianhua [WWR] as follows.

Let p be a prime number and a_{ij} $(1 \le i \le t,\ 1 \le j \le s)$ be a set of ts integers such that $p \nmid (a_{i1},\dots,a_{is})$ and $p \nmid (a_{1j},\dots,a_{tj})$ for all i and j. Denote by $p_1 = [\frac{p-1}{2}]$ and $p_2 = [\frac{p}{2}]$, where $[x]$ denotes the integer part of x. Consider the non-trivial solutions $(x_1,\dots,x_{s+t})$ and $(y_1,\dots,y_{s+t})$ of the system of linear congruences

$$a_{i1}x_1+\dots+a_{is}x_s + x_{s+i} \equiv 0 \pmod p, \quad 1 \le i \le t$$

and

$$a_{1j}y_1+\dots+a_{tj}y_t + x_{j+t} \equiv 0 \pmod p, \quad 1 \le j \le s,$$

where $-p_1 \le x_\nu \le p_2$, $-p_1 \le y_\nu \le p_2$, $1 \le \nu \le s+t$. Use q and Q to denote the minimum of $\bar{x}_1,\dots,\bar{x}_{s+t}$ and $\bar{y}_1,\dots,\bar{y}_{s+t}$ respectively. The following theorem is then true.

Theorem 1.2. q and Q satisfy

$$Q^{s+t-1} \le c(s,t)p^{(s+t)(s-1)}q.$$

This gives a generalization of a theorem due to Gel'fond which corresponds to $s = 1$ in Theorem 2 (cf. [Ko]). Further generalization is given by Wang and Ren [WR].

§2. Metrical theorems.

Let G_n denote the n-dimensional unit cube $0 \le x_1 < 1,\dots,0 \le x_n < 1$. Let A be a set in G_n. Denote by $|A|$ the measure of measurable set A. The following two theorems were proved by Wang and Yu.

Theorem 2.1. Let A_q $(q = 1,2,\dots)$ be a sequence of measurable sets in G_n. Suppose that $\psi(q) = |A_q|$ is a decreasing function of q. Let $N(h,\theta_1,\dots,\theta_n)$ be the number of integers satisfying $1 \le q \le h$ and $(\{q\theta_1\},\dots,\{q\theta_n\}) \in A_q$, where $\{x\}$ denotes the fractional part of x. Further, let $\Psi(h) = \sum_{q=1}^{h} \psi(q)$ and $\Omega(h) = \sum_{q=1}^{h} \psi(q)q^{-1}$. Then for almost all

$(\theta_1,\dots,\theta_n) \in G_n$, we have

$$N(h,\theta_1,\dots,\theta_n) = \Psi(h) + O(\Psi(h)^{1/2}\Omega(h)^{1/2}(\log \Psi(h))^{2+\varepsilon}).$$

Let $\bar{q} = (q_1,\dots,q_m)$ denote the lattice point with positive components, $\bar{q}\bar{\theta}$ the scalar product of $\bar{q}$ and $\bar{\theta}$ and $d(\bar{q}) = \sum_{\substack{d|q_i \\ 1\le i\le m}} 1$. We also use $\bar{q} \le h$ to denote $q = \max(q_1,\dots,q_m) \le h$.

Theorem 2.2. Let $\{A\bar{q}\}$ be a sequence of measurable sets in G_n. Put $\psi(q) = |A\bar{q}|$ and denote by $N(h,\bar{\theta}_1,\dots,\bar{\theta}_n)$ the number of lattice points $\bar{q}$ by satisfying $\bar{q} \le h$ and $(\{\bar{q}\bar{\theta}_1\},\dots,\{\bar{q}\bar{\theta}_n\}) \in A\bar{q}$, where $\bar{\theta}_i = (\theta_{i1},\dots,\theta_{im})$ $(1 \le i \le n)$. Put $\Psi(h) = \sum_{\bar{q}\le h} \psi(\bar{q})$ and $X(h) = \sum_{\bar{q}\le h} \psi(\bar{q})d(\bar{q})$. Then

$$N(h,\bar{\theta}_1,\dots,\bar{\theta}_n) = \Psi(h) + O(X(h)^{1/2}(\log X(h))^{3/2+\varepsilon})$$

holds for almost all $(\bar{\theta}_1,\dots,\bar{\theta}_n) \in G_{mn}$.

The proofs of theorems 2.1 and 2.2 are based on the method of Schmidt [Sc1]. Take Aq to be the set of points $(x_1,\dots,x_n)$ satisfying $0 \le x_i < \psi_i(q)$ $(1 \le i \le n)$ and suppose that $\prod_{i=1}^{n} \psi_i(q) = \psi(q)$ is a decreasing function of q in Theorem 2.1. Take $A\bar{q}$ to be the set such that $0 \le x_i < \psi_i(\bar{q})$ $(1 \le i \le n)$ in Theorem 2.2. Then we derive two theorems of Schmidt [Sc1]. Theorem 2.1. gives an improvement of a theorem of Gallagher [Gal1].

For any $\tau > 0$, let $E(\tau)$ be the set of points in G_n satisfying $0 \le x_i < 1/2$ $(1 \le i \le n)$ and $\tau x_1\cdots x_n < 1$. Let $f(1) = 1$ and $f(K) = K^{1+\varepsilon}$. Let $A\bar{q} = E(f(q_1),\dots,f(q_n))$, then it follows by Theorem 2.2 that the property A in Theorem 1.1 holds for almost all $(\theta_{11},\dots,\theta_{mn}) \in G_{mn}$. In other words, the set $E(m,n)$ of points $(\theta_{11},\dots,\theta_{mn})$ in $\mathbb{R}^{mn}$ for which the property A is false has Lebesgue measure zero. However, Yu [Y1] proved that

Theorem 2.3. The Hausdorff dimension of $E(m,n)$ is

$$mn - 1 + \frac{2}{2+\varepsilon}.$$

This is a generalization of a result due to Bovey and Dodson [BD]. Yu [Y2] also established the following theorem.

Theorem 2.4. For almost all real numbers θ there exist only finitely many positive integers q such that

$$q\|q\theta\|\,\|q\theta^2\| < (q\phi(q))^4,$$

where $\phi(q)$ is a given positive function such that $q\phi(q)$ is non-increasing and $\sum\phi(q)$ converges.

This confirms the following Baker's conjecture [Ba] for the case $n = 2$: that the inequality

$$q^{1+\varepsilon}\|q\theta\|\ldots\|q\theta^n\| < 1$$

has only finitely many solutions for almost all θ.

Remark. It was conjectured that the property A in Theorem 1.1 holds for $\theta_{ij} = e^{\gamma_{ij}}$ or $\theta_{ij} = \alpha_{ij}$ $(1 \le i \le m,\ 1 \le j \le n)$, where γ_{ij} and α_{ij} denote certain rational numbers and real algebraic numbers respectively (cf. [WY] and [Y1]).

§3. Linear forms in elliptic logarithms.

Let $\wp(z)$ be the Weierstrass elliptic function determined by the differential equation

$$(\wp'(z))^2 = 4(\wp(z))^3 - g_2\wp(z) - g_3,$$

where g_2, g_3 are algebraic numbers with $g_2^3 \neq 27g_3^2$. Let E be the elliptic curve defined by the equaton

$$y^2 = 4x^3 - g_2x - g_3.$$

We parameterize E as

$$x = \wp(z), \quad y = \wp'(z)$$

and write $p = p(z) = (\wp(z), \wp'(z))$, where the parameter z ranges over all complex numbers, not lying on the period lattice $\mathcal{L}$ of $\wp(z)$. We can view E as a group variety whose origin O is the point at infinity. We define

$$\Lambda = \beta_1u_1 + \ldots + \beta_nu_n \quad \text{with } n \ge 2,$$

where $\beta_1, \ldots, \beta_n$ are algebraic numbers, not all zero, and $u_1, \ldots, u_n$ are algebraic points of $\wp(z)$, by an algebraic point of $\wp(z)$ we mean a number $u \in \mathbb{C}$ such that either $\wp(u)$ is algebraic or u is a pole of $\wp(z)$. Assume further that $\wp(z)$ has complex multiplication over the complex quadratic field κ, and that $\mathbb{K}$ is a number field of degree D over $\mathbb{Q}$ containing the field κ and the numbers $g_2, g_3, \beta_1, \ldots, \beta_n, \wp(u_i), \wp'(u_i)$ with $1 \le i \le n$ and $u_i \notin \mathcal{L}$. We use the logarithmic absolute height $h(\alpha)$ for

$\alpha \in \bar{\mathbb{Q}}$ and $h(p)$ on the elliptic curve E. Suppose that $\beta_1,\dots,\beta_n$ are linearly independent over κ and $u_i \neq 0$ $(1 \le i \le n)$. For each i with $1 \le i \le n$, let $p_i = p(u_i)$ if $u_i \notin \mathcal{L}$ and let $p_i = 0$ if $u_i \in \mathcal{L}$. Suppose that V_i, V, Y, W are positive numbers satisfying

$$V_i \ge \max(1, h(p_i), |u_i|^2/D), \quad (1 \le i \le n),$$

$$V = \max_{1\le i\le n} V_i,$$

$$1 \le Y \le \min_{1\le i\le n} \log(eDV_i/|u_i|^2),$$

$$W = \max_{1\le i\le n} h(\beta_i).$$

Then the following two theorems were proved by Yu [Y3].

<u>Theorem 3.1</u>. There exists a constant $c > 0$ depending effectively on n and E such that

$$|\Lambda| > \exp(-cD^{n^2} V_1,\dots,V_n(W+\log(DV)+Y)^{n(n-1)}Y^{1-n^2}).$$

Moreover, if we suppose that

$$u_i \in \Pi = \{t_1w_1 + t_2w_2 | 0 \le t_1, t_2 < 1\}, \quad (1 \le i \le n),$$

where w_1, w_2 is a fundamental pair of periods of $\wp(z)$, and let A_i $(1 \le i \le n)$, B be positive numbers such that

$$A_i \ge \max(H(\wp(u_i)), e^e), \quad (1 \le i \le n),$$

$$A_1 \le A_2 \dots \le A_n,$$

$$B \ge \max(H(\beta_1),\dots,H(\beta_n), e),$$

where $H(\alpha)$ denotes the usual height of algebraic number α, then Theorem 3.1 has the following

<u>Corollary</u>. There exists a constant $c_1 > 0$ depending effectively on n and E such that

$$|\Lambda| > \exp(-c_1D^{n^2} \log A_1 \dots \log A_n(\log B + \log\log A_n)^{n(n-1)}(\log\log A_1)^{1-n^2}). \tag{3.1}$$

<u>Remark 1</u>. It may be interesting to notice that the product $\log A_1 \dots \log A_n$

in (3.1) is of exactly the same form as in the lower bounds for linear forms in logarithms of algebraic numbers (see Baker [Ba]). Thus the estimates are better than all previous estimates with respect to some of the parameters that appear.

Remark 2. Theorem 3.1 implies Masser's important theorem in the case $n = 2$ by simple argument from linear algebra.

Let D be a positive number such that

$$D \ge \max(3, [\kappa(g_2, g_3, \wp(u_i), \wp'(u_i); \quad 1 \le i \le n, \ u_i \notin \mathcal{L}) : \kappa(g_2, g_3)]);$$

denote by $\mathcal{O}$ the ring of endomorphisms of $\mathcal{L}$. The following theorem then holds.

Theorem 3.2. Suppose that $u_1, \ldots, u_n$ are linearly dependent over κ. Then there exist $\sigma_1, \ldots, \sigma_n$ in $\mathcal{O}$, not all zero, and a constant $c_2 > 0$ depending effectively on E, such that

$$\sigma_1 u_1 + \ldots + \sigma_n u_n = 0$$

and

$$|\sigma_i|^2 \le (c_2 (n-1)^2 D^3 (\log D)^2)^{n-1} V_1 \ldots V_n / V_i, \quad (1 \le i \le n).$$

Furthermore, if we suppose that

$$0 \ne u_i \in \Pi \quad (1 \le i \le n),$$

and let A_i $(1 \le i \le m)$ and D be positive numbers such that

$$A_i \ge \max(H(\wp(u_i), e))$$

$$D \ge \max(e^e, [\kappa(g_2, g_3, \wp(u_i), \wp'(u_i); 1 \le i \le n) : \kappa(g_2, g_3)]),$$

then we have

Theorem 3.2′ (Yu [Y3]). Under the hypothesis of Theorem 3.2, there exists a constant $c_2' > 0$ depending effectively on E such that

$$\sigma_1 u_1 + \ldots + \sigma_n u_n = 0$$

and

$$|\sigma_i|^2 \le (c_2' (n-1)^2 D^2 (\log D)^2)^{n-1} \log A_1 \ldots \log A_n / \log A_i, \quad (1 \le i \le n).$$

Theorems 3.2 and 3.2′ improve Anderson's results [An]. To obtain above Theorems 3.1 and 3.2, Yu established a zero estimate on the group variety $G_a^n \times E (\mathbb{C}^n \times E)$, which is sharper than that derived from the general results in

Masser and Wüstholz [MW].

§4. Algebraic independence.

In a series of papers Zhu [Z 1-6,8,9] investigated the transcendence and the algebraic independence in values of certain gap series and p-adic analogues. A main result is as follows.

Let $s, t \geq 1$ and

$$f_\nu(z) = \sum_{k=1}^{\infty} a_{\nu k} z^{\lambda_{\nu k}}, \quad (\nu = 1,\dots,s)$$

be s power series satisfying the following conditions:

(i) $\lambda_{\nu k}$ $(\nu = 1,\dots,s,\ k = 1,2,.1.)$ are s strictly monotone increasing sequences of natural numbers; (ii) the coefficients $a_{\nu k} \in \bar{\mathbb{Q}}^*$; and (iii) $f_\nu(z)$ $(\nu = 1,\dots,s)$ have the positive radii of covergence R_ν $(\nu = 1,\dots,2)$.

Let $D_n = [\mathbb{Q}(a_{11},\dots,a_{1n},\dots,a_{s1},\dots,a_{sn}) : \mathbb{Q}]$ and $A_n = \max(\overline{|a_{11}|},\dots,\overline{|a_{1n}|},\dots,\overline{|a_{s1}|},\dots,\overline{|a_{sn}|})$ where $\overline{|\alpha|} = \max(1, \max_{1\leq i\leq d}|\alpha^{(i)}|)$, $\alpha^{(1)} = \alpha$, $\alpha^{(i)}$ $(2 \leq i \leq d)$ are the conjugates of α. Denote by M_n the least common denominator of $a_{\mu\nu}$ $(1 \leq \mu \leq s,\ 1 \leq \nu \leq n)$.

Theorem 4.1 (Zhu [Z6]). Let $\theta_1,\dots,\theta_t \in \bar{\mathbb{Q}}$ with $0 < |\theta_1| < \dots < |\theta_t| < \min_{1\leq\nu\leq s} R_\nu$. If

$$\lim_{n\to\infty} \lambda_{\nu n}/\lambda_{\mu n} = 0 \quad (1 \leq \nu < \mu \leq s),$$

and

$$\lim_{n\to\infty}(\lambda_{sn} + \log M_n + \log A_n)/\lambda_{1,n+1} = 0,$$

then $f_\nu(\theta_\mu)$ $(1 \leq \nu \leq s,\ 1 \leq \mu \leq t)$ are algebraically independent.

This theorem extends and generalizes the earlier works of Adams [Ad], Knesser [Kn] et al.

In 1949 Gel'fond [Ge] established the well known "Gel'fond criteria of transcendence." In order to prove algebraic independence of numbers, one must generalize this criteria to the case of several variables. Various generalized forms have been obtained in recent years. Waldschmidt and Zhu [WZ] proved the following theorem.

Theorem 4.2. Let $n \geq 1$ and $(\theta_1,\dots,\theta_n) \in \mathbb{C}^n$. Denote by $A(\theta_1,\dots,\theta_n)$ a set of real numbers η (> 1) which satisfies the following conditions:

there is a constant $N_0 > 0$ such that for every real number $N > N_0$ and every point $z = (z_1,\dots,z_n) \in \mathbb{C}^n$ satisfying $|\theta_i - z_i| < \exp(-2N^{\eta})$ $(1 \le i \le n)$, there is a non-zero polynomial $F(z_1,\dots,z_n) \in \mathbb{Z}[z_1,\dots,z_n]$ with $t(F) \le N$ and $0 < |F(\theta_1,z_2,\dots,z_n)| < \exp(-N^{\eta})$, where $t(F) = \max(\log H(F), 1+\deg(F))$. Define

$$J(\theta_1,\dots,\theta_n) = \begin{cases} 1, & \text{if } A(\theta_1,\dots,\theta_n) = \emptyset, \\ \sup A(\theta_1,\dots,\theta_n), & \text{if otherwise.} \end{cases}$$

Then we have

$$J(\theta_1,\dots,\theta_n) \le 2^n.$$

Recently this theorem was improved by Zhu [Z7] to

$$J(\theta_1,\dots,\theta_n) \le (2+\sqrt{3})2^{n-2} \quad (n \ge 2).$$

In his other papers, Zhu [Z4,9,11] obtained an improvement in the criteria of algebraic independence due to Durand [D].

Let $s \ge 1$, $z = (z_1,\dots,z_s)$ and Ω be the completion of $\bar{\mathbb{Q}}_v$. Denote by $|\ |_v$ the normalized v-adic valuation. Let $p(z_1,\dots,z_s) \not\equiv 0 \in \mathbb{Z}[z_1,\dots,z_s]$, $\partial(P) = \sum_{i=1}^{s} \deg_{z_i}(P)$ and $\Lambda(P) = 2^{\partial(P)}L(P)$, where $L(P)$ is the length of $P(z_1,\dots,z_s)$. We define the order function for $\theta = (\theta_1,\dots,\theta_s) \in \Omega^s$ as follows:

$$O(u|\theta) = \sup \log|P(\theta_1,\dots,\theta_s)|_v^{-1},$$

where the sup is taken over all $P(z_1,\dots,z_s) \not\equiv 0 \in \mathbb{Z}[z_1,\dots,z_s]$ with $\Lambda(P) \le u \in \mathbb{N}$. Let M be a finite or infinite set of real numbers. Denote by θ_M a set of $\theta_\mu \in \Omega$ $(\mu \in M)$ and by T a subset of M. Let $\{\theta_{\tau n}\}$ be an infinite sequence from Ω for $\tau \in T$ and $\{\theta_{Tn}\}$ be a set constituted by $\{\theta_{\tau n}\}$ $(\tau \in T)$.

<u>Theorem 4.3</u> (Zhu [Z11]). Suppose that $\theta_M \in \Omega$ with $M \ne \emptyset$. If for every finite subset $T \ne \emptyset$ from M there exists a set of infinite sequence $\{\theta_{Tn}\} \in \Omega$ and an infinite sequence $\{u_n\} \in \mathbb{N}$ $(u_n \to \infty)$ such that

(i) $\lim_{n\to\infty} |\theta_{\tau n}|_v = |\theta_\tau|_v$, $|\theta_{\tau n} - \theta_\tau|_v > 0$ $(n \ge 1)$ for every $\tau \in T$;

(ii) $\sum_{\tau\in T} |\theta_{\tau n} - \theta_\tau|_v \sim \max_{\tau\in T} |\theta_{\tau n} - \theta_\tau|_v$ $(n \to \infty)$;

and

(iii) $\max_{\tau\in T} |\theta_{\tau n} - \theta_\tau|_v \le \gamma \exp(-O(u_n|\theta_{Tn}))$, $(n \ge n_0)$,

where the constants γ and n_0 depend only on T, then the θ_M are algebraically independent.

§5. Transcendence of numbers of certain classes.

Mahler [Mah1] proved that the decimal fraction

$$0.1234567891011\ldots$$

is transcendental. Later on, he investigated a more general problem [Mah3]. Let $a = \{a_1, a_2, a_3, \ldots\}$ be a fixed sequence of positive integers. Let

$$\sigma(a) = 0.d_1d_2d_3\ldots$$

be the expansion to the basis $g \geq 2$ in which we write successively after the point the expansions to the basis g of the integers

1, a_1 times repeated,

2, a_2 times repeated,

3, a_3 times repeated,

etc., thus $d_1, d_2, d_3, \ldots$ are the resulting digits of $\sigma(a)$. Mahler proved that if $\alpha(n)$ is an arbitrary positive integral-valued arithmetic function for $n = 1,2,\ldots$, and if $a_k = \alpha(n)$ for $g^{n-1} \leq k \leq g^n - 1$, then $\sigma(a)$ is transcendental. More generally, a sufficient condition for transcendency of $\sigma(a)$ is obtained by Zhu [Z1].

Theorem 5.1. Let $\{n_1, n_2, \ldots\}$ and $\{k_1, k_2, \ldots\}$ be two increasing sequences of natural numbers satisfying $g^{n_\nu} - 1 \leq k_\nu \leq g^{g_\nu + 1} - 2$ $(\nu \geq 1)$ and if

$$\underline{\lim}_{\nu\to\infty}(E_{n_\nu}^{(k_\nu+1)} / E_{n_\nu}^{(k_\nu)}) > 1,$$

then $\sigma(a)$ is a transcendental number where $E_{n_\nu}^{(k_\nu)}$ is defined in terms of a_ν, n_ν, k_ν, g which is explicitly given.

The proof of Theorem 5.1 is based on a p-adic generalization of Roth's theorem due to Ridout [R].

As an application of Schmidt's theorem [Sc2] on simultaneous approximation to algebraic numbers by rationals we gave a criteria of transcendence and some transcendental numbers. Let $q_n = q_n(\alpha)$ denote the denominator of the n-th convergent for the real number α. Suppose that a_n, b_n, c_n ($n = 1,2,\ldots$) are three sequences of natural numbers satisfying the following conditions:

$$\lim_{n\to\infty} c_{n+1}/c_n = c, \quad \text{where } c \text{ is a constant}$$

$$b_n | b_{n+1}, \quad \log b_n = o(c_n) \quad (n \to \infty),$$

$$a_n = o(q_{c_n}^{\varepsilon}), \quad (n \to \infty, \quad \text{for any} \quad \varepsilon > 0),$$

$$\lim_{n\to\infty} \frac{a_n}{b_n} = \lambda = 0, \infty \quad \text{or irrational number.}$$

Denote

$$\xi = \sum_{n=1}^{\infty} \frac{1}{q_{c_n}}, \quad \xi_\lambda = \sum_{n=1}^{\infty} \frac{a_n}{b_n q_{c_n}}.$$

Zhu, Wang, and Xu [ZWX] obtained

Theorem 5.2. If an algebraic number α satisfies

$$\lim_{n\to\infty} \sqrt[n]{q_n} = \beta > 1, \quad \text{where } \beta \text{ is a constant and } \quad q_n^{-1} \ll \|\alpha q_n\| \ll q_n^{-1},$$

then (i) ξ_λ is a transcendental number for $c > 1+\sqrt{3}$; (ii) one at least of ξ, ξ_λ is transcendental number for $c > 1 + \sqrt{2}$.

An algorithm of p-adic simple continued fractions was introduced by Wang Lianxiang [W4]. Wang obtained a sufficient condition for certain p-adic transcendental numbers whose sum, difference, product and quotient are all p-adic transcendental numbers by using their expansions into p-adic simple continued fractions. Further, Wang and Mo [WM] established the transcendence for the value ξ^{η} of power exponential function in p-adic numbers ξ and η under some conditions; this result is a p-adic analogue of a theorem due to Bundshuh [Bu].

§6. E- and G-functions.

In his fundamental paper in 1929, Siegel [Si] developed a method for studying the arithmetic properties in the values of certain classes of analytic functions known as E- and G-functions. In 1954, the Siegel-Shidlovsky theorem on E-functions was established by Shidlovsky. An excellent account of this theorem was given by Mahler in his book [Mah4] in 1976. By following Siegel's method, we considered lower bounds for linear forms and polynomials in the values of a class of E- and G-functions.

Let $\mathbb{K}$ be an algebraic number field of degree d over $\mathbb{Q}$, and let $O_{\mathbb{K}}$ denote the domain of integers in $\mathbb{K}$. For every place ν of $\mathbb{K}$ we normalize the absolute value $\| \|_\nu$ so that $|p|_\nu = p^{-d_\nu/d}$ if $\nu | p$ and $|x|_\nu =$

$|x|^{-d_\nu/d}$ if $\nu|\infty$, where $d_\nu = [\mathbb{K}_\nu : \mathbb{Q}_\nu]$ and $|\ |$ denotes the absolute value in $\mathbb{R}$ or $\mathbb{C}$. We use α_ν to denote 1 if $\nu|p$, and d_ν/d if $\nu|\infty$. Let $c \geq 1$ be a constant, and $a_0, a_1, \ldots$ be a sequence of elements of $\mathbb{K}$ satisfying $|a_m|_\nu \leq c^{\alpha_\nu m}$ for all $\nu|\infty$ and $m \geq 0$ and so that there is a sequence of natural numbers $q_0, q_1, \ldots$ with $q_m \leq c^m$, $m \geq 0$, such that $q_m a_i \in O_{\mathbb{K}}$, $0 \leq i \leq m$, $m \geq 0$. Then the E-function

$$f(z) = \sum_{m=0}^{\infty} a_m z^m/m!$$

is said to belong to the class $\mathbb{K}E(c)$, and the G-function

$$g(z) = \sum_{m=0}^{\infty} a_m z^m$$

is said to belong the the class $\mathbb{K}G(c)$.

Let $\{f_{ij}(z)\}$ $(1 \leq i \leq k,\ 1 \leq j \leq n_i)$ belong to the class $\mathbb{K}E(c)$. Suppose that the $\{f_{ij}(z)\}$ and 1 are linearly independent over $\mathbb{C}(z)$ and satisfy the following system of differential equations

$$y'_{ij} = Q_{ij0}(z) + \sum_{m=0}^{n_i} Q_{ij\ell}(z) y_{i\ell}, \quad (1 \leq i \leq k,\quad 1 \leq j \leq n_i), \tag{6.1}$$

where $Q_{ij\ell}(z) \in \mathbb{K}(z)$. We denote the linear form in E-functions by

$$L(z) = x_{01} + \sum_{i=1}^{k} \sum_{j=1}^{n_i} x_{ij} f_{ij}(z),$$

where $x_{ij} \in O_{\mathbb{K}}$, not all zero. Suppose that $\alpha \in \mathbb{K}$ is non-zero and different from the poles of all $Q_{ij\ell}(z)$. By L_ν we mean a linear form obtained by considering $L(\alpha)$ in the corresponding completion $\mathbb{K}_\nu$. Xu [X2] proved that

<u>Theorem 6.1.</u> Put $d \geq 2$. There are explicit constants c_1 and c_2 depending only on the functions $f_{ij}(z)$, $\mathbb{K}$ and (6.1) such that

$$\max_{\nu|\infty} |L_\nu|_\nu^{d/d_\nu} > \left(\prod_{i=1}^{K} x_i^{-n_i} \right) x^{-c_1/(\log\log x)^{1/2}}$$

provided $x > c_2$, where $x_i = \max\limits_{1 \leq j \leq n_i} (\max\limits_{\nu|\infty} |x'_{ij}|_\nu^{d/d_\nu})$,

$$x = \max_{1<i\le k}\left(x_i, \max_{\nu|\infty}|x_{01}|_\nu^{d/d_\nu}\right).$$

Remark. In the case $d = 1$ we have given the proof of Theorem 6.1 in [Xw]; it is a generalization of the result due to Makarov [Mak] and of our earlier theorem [XY] in the case $n_i = 1$. Similarly, under the hypotheses of Galochkin's conditions [Gal0], Xu [X1,2] and Wang [W5] gave lower bounds for linear forms concerning G-functions in the archimedian and p-adic cases, respectively.

Using very interesting new ideas, Chudnovsky [C] recently succeeded in considering the arithmetic properties of the values of classical G-functions at certain rational points without the additional Galochkin's conditions. A similar and sometimes stronger result may be found in Osgood [O]. We shall give below a generalization of these results to algebraic number fields in both the archimedian and p-adic case. Our proof is based on the use of Padé approximation of the second kind and on local to global techniques. Let $\{g_i(z)\}$ $(1 \le i \le n)$ belong to $\mathbb{K}G(c)$. Suppose that the $\{g_i(z)\}$ and 1 are linearly independent over $\mathbb{K}(z)$ and satisfy the system of differential equations

$$y_i' = \sum_{j=1}^{n} Q_{ij}(z)y_j, \quad (1 \le i \le n) \tag{6.2}$$

where $Q_{ij}(z) \in \mathbb{K}(z)$. The linear form $\ell(z)$ is denoted by

$$\ell(z) = H_0 + \sum_{i=1}^{n} H_i g_i(z),$$

where $H_i \in \mathbb{K}$, not all zero. By $h(\theta)$ and $h(H)$ denote the absolute height for $\theta \in \mathbb{K}$ and the vector $H = (H_1,\dots,H_n) \in \mathbb{K}^n$. Then we have

Theorem 6.2 (Väänänen and Xu [VX1]). Let u and ε, $0 < u, \varepsilon < 1$, be given. There then exists an explicit constant λ (> 0) depending only on u,ε and $\{g_i(z)\}$ such that if $\theta \in \mathbb{K}$ is non-zero and different from the poles of all $Q_{ij}(z)$, and satisfies $\log h(\theta) > \lambda$, and $\log|\theta|_\nu \le \min\left[\left(\frac{n\varepsilon}{(n+1)(n+\varepsilon)} - 1\right)\log h(\theta), -\alpha_\nu \log 2c\right]$, then we have

$$\log|\ell_\nu(\theta)|_\nu > -(n+1+\varepsilon)\log h(H) + \log^+\max_i |H_i|_\nu$$

for all $h(H) > c_0$, where c_0 is a positive constant depending on u,ε,θ, and (6.2). As usual: $\log^+ a \equiv \log\max(1,a)$ for $a \ge 0$.

We further consider the algebraic independence in the values of

G-functions in both the archimedian and p-adic case in [VX2].

Under the hypotheses of Galochkin's conditions, Wang [W1,2,3] obtained simultaneous approximations for some numbers related to G-functions and introduced a p-adic F-function and discussed the arithmetic properties of values of p-adic F-functions.

References

[Ad] Adams, W., On the algebraic independence of certain Liouville numbers, J. Pure. Appl. Algebra, B (1978), 41-47.

[An] Anderson, M., Linear forms in algebraic points of an elliptic function, Ph.D. thesis, Univ. Nottingham, 1978.

[Ba] Baker, A., Transcendental number theory, Camb. Univ. Press, 1978.

[BD] Bovey, J.D. & Dodson, M.M., The fractional dimension of sets whose simultaneous rational approximations, Bull. London Math. Soc., 10 (1978), 213-218.

[Bu] Bundschuh, P., Transcendental Continued Fractions, J. Number Theory, 18 (1984), 91-98.

[C] Chudnovsky, G.V., On applications of diophantine approximations, Proc. Natl. Acad. Sci. USA, 8 (1984), 7261-7265.

[D] Durand, M.A., Independance algébrique de nombres complexes et critère de transcendance, Comp. Math. 35 (1977), 259-267.

[Gall] Gallagher, P.X., Metric simultaneous diophantine approximation, JLMS, 37 (1962), 387-390.

[Galo] Galochkin, A.I., Lower bounds for polynomials of the values of the class of analytic functions, Math. Sb. 95 (137), (3), (1974), 396-417.

[Ge] Gel'fond, A.O., Transcendental and Algebraic Numbers, Moscow, 1952.

[Ko] Korobov, N.M., Certain questions of the theory of diophantine approximations, Uspehi Mat. Nauk, SSSR, 135 (1967), 83-118.

[Kn] Knesser, H., Eine Kontinnumsmächtige algebraisch unabhängige Menge reeller Zahlen, Bull. Sco. Math. Belg., 12 (1960), 23-27.

[Mah1] Mahler, K., Über die Dezimalbruchentwicklung gewisser Irrationalzablen, Mathematica (Zutphen) 4B (1937).

[Mah2] __________, Ein Übertragungsprinzip für Linearformen, Časopis 68 (1939), 82-92.

[Mah3] __________, On a class of transcendental decimal fractions, Comm. Pure Appl. Math. 29 (1976), 717-725.

[Mah4] __________, Lectures on transcendental numbers. Lecture Notes in Math 546, Springer-Verlag, Berlin-New York, 1976.

[Mak] Makarov, N.Yu., Estimates of the measure of linear independence for the values of E-functions, Vestnik Moskow, Univ. Ser. 1 Mat. 2 (1978), 3-12.

[MW] Masser, D.M. & Wüstholz, G., Zero estimates on group varieties, I, Invent. Math. 64 (1981), 489-516.

[O] Osgood, C., Product Type Bounds on the Approximation of Values of E and G Functions, Monatshefle für Mathematik, 102 (1986), 7-25.

[R] Ridout, D., Rational approximations to algebraic numbers, Mathematika 4 (1957), 125-131.

[Sc1] Schmidt, W.M., A metrical theorme in diophantine approximation, Can. J. Math., 12 (1960), 619-630.

[Sc2] __________, Simultaneous approximation to algebraic numbers by rationals, Acta Math., 125 (1970), 189-201.

[Si] Siegel, S.L. Über einige Anwendungen diophantischer Approximationen, Abh. Preuss. Akad. Wiss, 1 (1929).

[SW] Schmidt, W.M. & Wang Yuan, A note on a transference theorem of linear forms, Sci. Sin. 22 (1979), 276-280.

[VX1] Väänänen, K. & Xu Guangshan, On linear forms of G-functions, Acta Math, to appear.

[VX2] __________, On arithmetic properties of the values of G-functions, J. Austral Math. Soc. Ser. A, to appear.

[W1] Wang, Lianxiang, On p-adic E-functions and G-functions, Sci. Sin. Ser. A, 26 (1983), 1266-1274.

[W2] __________, On p-adic F-functions, J. Austral. Math. Soc. Ser. A, 35 (1983, 269-280.

[W3] __________, On p-adic F-function with their rank, Sci. Sin. Ser. A, 27 (1984), 709-219.

[W4] __________, p-adic continued fractions (I), Sci, Sin. Ser. A, 28 (1985), 1009-1017; (II), ibid., 1018-1029.

[W5] __________, On lower bounds for linear forms in the values of a class of p-adic G-functions, Acta Math. Sin. New Ser., to appear.

[Wy] Wang Yuan, Number theoretic method in numerical analysis (in this monograph).

[WM] Wang Lianxiang & Mo Detze, p-adic continued fractions (III), Acta Math. Sin. Nwe Ser., to appear.

[WR] Wang Lianxiang & Ren Jianhua, On a transference theorem of the systems of linear congruence mod p^n, Kexue Tongbao, Sin. Special Issue on Math. Phys. and Chem., 1980, 127-130.

[WW] Wang Yuan & Wang Lianxiang, Diophantine approximations in China, Review of Chinese Math., 1 (1985), 160-171.

[WWR] Wang Yuan, Wang Lianxiang & Ren Jianhua, A note on a transference theorem of the systems of linear congruences, Acta Math. Sin. 24 (1981), 303-307.

[WY] Wang Yuan & Yu Kunrui, A note on some metrical theorems in diophantine approximation, IHES/M, 297 (1979); Chinese Ann. Math. 2 (1981), 1-12.

[WYZ] Wang Yuan, Yu Kunrui & Zhu Yaochen, On a transference theorem on linear forms, Acta Math. Sin., 22 (1978), 237-240 (Chinese).

[WZ] Waldschmidt, M. & Zhu Yaochen, Une généralization en plusieurs variables d'un critère de transcendance de Gel'fond, C.R. Acad. Sci. Paris, Ser. I, t 297 (1983), 229-232.

[X1] Xu Guangshan, Lower estimates of linear forms of a class of G-functions, Acta Math. Sin., 24 (1981), 578-586 (Chinese).

[X2] __________, A note on linear forms in a class of E-functions and G-functions, J. Austral. math. Soc. Ser. A, 35 (1983), 338-348.

[X3] __________, On effective results of simultaneous approximations to certain algebraic numbers, Sci. Sin. Ser. A, 29 (1986), 149-164.

[XW] Xu Guangshan & Wang Lianxiang, On explicit estimates for linear forms in the values of a class of E-functions, Bull. Austral. Math. Soc., 25 (1982), 37-68.

[XY] Xu Guangshan & Yu Kunrui, On some diophantine inequalities involving a class of Siegel's E-functions, Kexue Tongbao, Sin. 24 (1979), 481-486 (Chinese).

[Y1] Yu Kunrui, Hausdorff dimension and simulatneous rational approximation, J. London Math. Soc., 24 (1981), 79-84.

[Y2] __________, A note on problem of Baker in metrical number theory, Math. Proc. Camb. Phil. Soc., 90 (1981), 215-227.

[Y3] __________, Liner forms in elliptic logarithms, J. of Number Theory, 20 (1985), 1-69.

[YY] Yu Xiuyuan and Yu Kunrui, A class of linear forms in complex logarithms, Kexue Tongbao, Sin. 25 (1980), 580-582.

[Z1] Zhu Yaochen, A note on the transcendence of Mahler's decimal fraction, Acta Math. Sin. 24 (1981), 247-253 (Chinese).

[Z2] __________, Algebraic independence of the values of certain gap series in rational points, Acta Math. Sin., 25 (1982), 333-339 (Chinese).

[Z3] __________, An infinite system of algebraically independent numbers, Kexue Tongbao, Sin. 27 (1980), 251-253.

[Z4] __________, On algebraic independence of certain power series of algebraic numbers, Kexue Tongbao, Sin. 28 (1983), 427.

[Z5] __________, A theorem about algebraic independence and its application, J. of Math. Res. and Exp. 3 (1983), 1-4.

[Z6] __________, On the algebraic independence of certain power series of algebraic numbers, Chin. Ann. of Math. Ser. B, 5 (1984), 109-117.

[Z7] __________, A generalization in several variables of a transcendence criterion of Gel'fond (II), C.R. Math. Rep. Acad. Sci. Canada, 6:5 (1984), 297-302; Publications Math. Paris VI, Problémes Diophantiens 1983-1984, 6.1-6.16.

[Z8] __________, Algebraic independence property of values of certain gap series, Kexue Tongbao, Sin, 29 (1984), 1409-1412.

[Z9] __________, Criteria of the algebraic independence of complex numbers, Kexue Tongbao, Sin. 30 (1985), 973-975.

[Z10] __________, Arithmetic properties of gap series with algebraic coefficients, Acta Arith., to appear.

[Z11] __________, On a criterion of algebraic independence of numbers, Kexue Tongbao, Sin, to appear.

[ZR] Zhu Yaochen & Ren Jianhua, The approximations to e and e^{π} by algebraic numbers, J. of Northwest Univ. No. 1 (1982), 29-42 (Chinese).

[ZWX] Zhu Yaochen, Wang Lianxiang & Xu Guangshan, On the transcendence of a class of series - an application of Schmidt's theorem, Kexue Tongbao, Sin., 25 (1980), 1-6.

Institute of Mathematics
Academia Sinica, Beijing

Contemporary Mathematics
Volume **77**, 1988

QUADRATIC FORMS AND HERMITIAN FORMS

Li Delang and Lu Hongwen

Chinese mathematicians have paid much attention to the theory of quadratic forms. It is also one of the earliest and most fruitful areas in number theory developed in China. Professor Ko and students have especially studied quadratic forms for about 50 years and have published many papers. Since 1980, Dai (with T.Y. Lam) and Peng have done extensive work with the notion of level in algebra and topology. The analytic theory of Hermitian forms has also been studied. This article will survey some of the main results in these areas obtained by Chinese mathematicians.

1. Representation of positive definite forms as a sum of linear forms

Let $f = \sum_{i,j=1}^{n} a_{ij}x_ix_j$, $a_{ij} \in \mathbb{Z}$ be a positive definite form. Let R_n (R'_n resp.) be the smallest integer r_n such that, for every n-ary form f, there exist r_n linear forms $L_i(x_1,\dots,x_n) = \sum_i b_{ij}x_j$, where $b_{ij} \in \mathbb{Q}$ ($\mathbb{Z}$ resp.) such that

$$f = \sum_{i=1}^{n} L_i(x_1,\dots,x_n)^2.$$

The effort for finding R_n and R'_n began in 1904 when Landau proved $R_2 \le 5$. In 1937 Mordell [45] proved $R'_2 = 5$ and $R_n \le n+3$. In 1937, Ko [15] gave a simpler proof of Mordell's result $R_n \le n+3$ and proved that $R'_n = n+3$ for $n = 2,3,4,5$. In 1938, Ko [16] considered an example

$$f_n = \sum_{i=1}^{n-1} x_i^2 + bx_n^2$$

where $b = 4^{\alpha}(8\beta+7)$, α,β non-negative integers. He proved that f_n cannot be expressed as a sum of squares of $n+2$ linear forms. It follows that $R_n = n+3$. In 1957, Ko gave another form

$$g_n = \sum_{i=1}^{n-3} x_i^2 + 3x_{n-2}^2 + 3x_{n-1}^2 + 2x_n^2$$

with determinant 18 which cannot be expressed as a sum of squares of $n+2$ linear forms. In the same paper he found a necessary and sufficient condition that a ternary form can be expressed as a sum of squares of 4 linear forms of rational coefficients.

Ko [14] also considered the representation of indefinite forms as algebraic sum of squares of linear forms. He proved that any binary form $f = \sum_{i,j=1}^{2} a_{ij}x_ix_j$ can be represented in the form

$$f = \sum_{i=1}^{5} \varepsilon_i(b_{i1}x_1+b_{i2}x_2)^2 \tag{1}$$

where $b_{ij} \in \mathbb{Z}$ and $\varepsilon_i = \pm 1$. Furthermore, for any given $\varepsilon_i = \pm 1$, expression (1) is possible except some trivial exceptional cases (for example f is positive definite and $\varepsilon_i = -1$). He also considered the possibility of representation

$$f(x_1,x_2) = \sum_{i=1}^{4} \varepsilon_i(b_{i1}x_1+b_{i2}x_2)^2 \tag{2}$$

and proved that the representation (2) is impossible when $a_{11} \equiv a_{22} \equiv 2 \pmod 4$, a_{12} odd, $\varepsilon_1 = \varepsilon_2 = 1$, $\varepsilon_3 = \varepsilon_4 = -1$, and (2) is possible in other cases.

In 1940, Ko [20] proved that for n-ary unimodular form f, it is always possible to choose $\varepsilon_i = \pm 1$ and L_i such that

$$f = \sum_{i=1}^{n+3} \varepsilon_i L_i(x_1,\dots,x_2)^2.$$

In the same paper he gave a new proof for Meyer's theorem that indefinite forms in 5 or more variables are isotropic.

2. Non-decomposable forms

Let f be a positive definite integral form. We say f is non-decomposable if f cannot be expressed as a sum of two non-negative integral forms.

In 1937, Mordell proved that there exists no non-decomposable form for $n \le 5$ and there exists such form for $n = 8$.

In 1938, Erdös and Ko [11] proved that there exists n-ary non-decomposable form for $n \ge 12$ except possibly $n = 13, 17, 19, 23$. In 1942, Ko [21],[22] proved that the only non-decomposable forms of 7 or 8 variables (not necessarily unimodular) are ones equivalent to f_7 (of det 2) and f_8 (det 1), where

$$f_n = \sum_{i=1}^{n} x_i^2 + \left[\sum_{i=1}^{n} x_i\right]^2 - 2x_1x_2 - 2x_1x_n.$$

In 1958, Ko [26] found out the answer for $n = 13$: there is no non-decomposable unimodular form of 13 variables.

Recently Zhu Fuzu [53],[54] gave unimodular non-decomposable forms for $n = 17, 19, 23$. Hence there exist such forms for $n = 8, 12$ and $n \ge 14$. Zhu also proves that for unimodular forms the indecomposability is equivalent to the non-decomposability.

3. Class number of definite unimodular forms

Let $C_{n,1}$ denote the class number of n-ary positive definite unimodular forms.

Hermite proved that $C_{n,1} = 1$ for $n \le 7$.

It was known to Korkine and Zolotareff (1873) that $C_{8,1} = 2$. Ko [17] proved in 1938 that only one class represents 1.

Mordell proved in 1938 that $C_{8,1} = 2$. Later Ko [19] proved $C_{n,1} = 2$ for $n = 9, 10, 11$.

Ko came back to this problem in 1958-60 and proved [26],[27],[28] that $C_{12,1} = C_{13,1} = 3$, $C_{14,1} = 4$, $C_{15,1} = 5$, and $C_{16,1} \ge 8$. Furthermore, he found out a representative form in each class considered.

4. Minimum of definite forms

Let f be a positive definite n-ary form with real coefficients. Let D be the discriminant of f and $\min f = \min_{x \in \mathbb{Z}} f(x_1, \ldots, x_n)$.

It is well known that there exists a constant C_n, depending only on n but not on f, such that

$$\frac{\min f}{D^{1/n}} \le C_n. \qquad (3)$$

Let γ_n be the smallest real number C_n satisfying (3) for all f.

The exact values of γ_n for $n \le 8$ have been known for a long time. Chaundy [2] proved $\gamma_9 \ge 2$, $\gamma_{10} \ge 2\sqrt[10]{4/3}$ in 1946. In fact, he claimed that the equalities held, but the proof was incomplete.

In 1963, Ko and Zheng Dexun [34] proved that $\gamma_{11} \ge 2\sqrt[11]{512/243}$.

5. Forms representing the same numbers

If f and g are equivalent forms, then they certainly represent the same numbers. But the converse is not true. Legendre asserted without proof that two positive definite binary forms representing the same numbers are equivalent. Unfortunately, this assertion is not correct. In 1895, Bauer proved Legendre's assertion for primitive forms with the same determinant. In 1938, Delone [9] proved by geometric consideration that two positive definite binary forms with real coefficients are equivalent if they represent the same numbers, with the sole exception of the pair equivalent to

$$x^2 + xy + y^2$$

$$x^2 + 3y^2$$

and their scalings.

Ko and Zheng Dexun (1978) [35] and Watson (1979) [50] gave two new proofs of Delone's result.

In 1982, Li Delang [41] considered the problem for the indefinite forms with rational coefficients. The result is as follows:

Let $f(x,y) = ax^2 + bxy + cy^2$ be a primitive binary form with discriminant D. If $D \equiv 5 \pmod 8$ and the equation $t^2 - Du^2 = 4$ has an integer solution with u odd, then $f(x,y)$ and $g(x,y) = f(x,2y)$ represent the same numbers. Moreover, by scaling such $f(x,y)$ and $g(x,y)$ we obtain all pairs of indefinite binary forms with rational coefficients which represent the same numbers but are not equivalent. The proof is based on local-global principle and hence does not work for forms with real or complex coefficients. Recently Li Delang [42] proved a generalization of this result for forms with complex coefficients.

If $n \ge 3$, there are several examples of pairs of n-ary forms representing the same numbers but that are not equivalent. The first example is:

$$f = x^2 + 3(y^2+yz+z^2)$$

$$g = x^2 + x + y^2 + 9z^2$$

given by Watson [49]. Timofeev [48] obtained more examples in 1963.

In 1978, Ko and Zheng Dexun [35] gave more examples for $n \geq 3$. later Ko and Zheng [37],[52] studied diagonal ternary forms representing the same numbers. Hisa [12] studied this problem and posed several conjectures. One of them says two positive ternary quadratic forms in the same genus are equivalent if they primitively represent the same integers.

6. Class number and representatives of positive definite forms

From 1959 to 1964, Ko and his students [30],[32],[33] gave three tables containing representatives of classes of positive definite integral forms of 4,5 and 6 variables and of $D \leq 50, 25$ and 25 respectively, where D denotes discriminant.

7. Witt ring of fields of finite square class number

Szymiczek [47] classified fields of square class number 8 with respect to their Witt ring in 1976. After recognizing that two cases (one for real fields and another for non-real fields) were missed, he posed an open problem in 1976 for finding the solutions. In 1980, Li Delang [40] gave a method of constructing fields with finite square class number and calculated their Witt rings. In particular, he obtained infinitely many real fields with Witt ring $\mathbb{Z}[x,y]/(x^2-1,y^2-1,2(x-1),2(y-1))$ and non-real fields with Witt ring $\mathbb{Z}_4[x,y]/(x^2-1,y^2-1,2(x-1),2(y-1))$ which answers Szymiczek's question. This problem was also solved by Kula and Szymiczek [38],[39] in a different way.

8. Witt vanishing theorem for quadratic forms over fields of characteristic 2

Witt [51] proved his famous vanishing theorem only for fields of characteristic $\neq 2$. C. Arf [1] generalized Witt theorem to fields of characteristic 2, but he considered only the case of genus $d = 0$. For the case of genus $d > 0$, Dai Zongduo [4] proved the Witt theorem as well as the transitivity theorem and the extension theorem in 1966.

9. Level in algebra and topology

Let A be a commutative ring. The level $S(A)$ of A is the smallest integer n such that -1 is a sum of n squares in A. If -1 is not a sum of squares in A, we define $S(A)$ to be ∞. Let $F \in A[X_1,\dots,X_m]$ be a form (i.e., a homogeneous polynomial) of degree d over A. We say that F is isotropic over A if F has a unimodular zero vector, i.e., if there exist $a_1,\dots,a_m \in A$ generating A as an ideal, such that $F(a_1,\dots,a_m) = 0$. The sublevel $\sigma(A)$ of A is defined by

$$1 \le \sigma(A) := \min\{n : (n+1)\langle 1\rangle \text{ is isotropic over } A\},$$

where $r\langle 1\rangle$ denotes the r-dimensional quadratic form $X_1^2 + \dots + X_r^2$ over A. If 2 is invertible in A, it is easy to see that $\sigma(A) = S(A)$ or $S(A)-1$.

By a well known theorem of A. Pfister, if A is a field with $S(A) < \infty$, then $S(A)$ must be a power of 2 (and any power of 2 is possible). Knebusch and Baeza proved a similar theorem for semilocal rings, but very little is known about levels of commutative rings in general. M. Knebusch [13] asks what type of integers can be the level of a ring (see also LNM V. 655, p. 184).

This problem was settled by Dai, Lam and Peng [6] in 1980. They proved that for any given $n \ge 1$, there exists an integral domain A with $S(A) = n$. Moreover, they proved for any $n \ge 1$, there exists an integral domain A with $(\sigma(A), S(A)) = (n,n)$, and for $n = 1,2,4,8$, there exists an integral domain A with $(\sigma(A), S(A)) = (n-1,n)$, but $S(A) = \sigma(A)$ if $S(A) = 1,2,4$ or 8.

Their result follows from a close relationship between the level in algebra and the level in topology. Let S^n be the n-dimensional real sphere with antipodal map as an involution. Let X and Y be two topological spaces with involutions ε and ε_1 respectively. An equivariant map $f : X \to Y$ is a continuous involution-preserving map. For any topological space X with involution ε, the following two invariants

$$S(X) = \inf\{n : \text{there exists an equivariant map } X \to S^{n-1}\},$$

and

$$S'(X) = \sup\{m : \text{there exists an equivariant map } S^{m-1} \to X\},$$

are called, respectively, the level and colevel of X. For any space (X,ε), let $A_{(X,\varepsilon)}$ (or A_X for short) be the ring of all complex-valued functions $f : X \to \mathbb{C}$ such that $f(\varepsilon x) = \overline{f(x)}$ for any $x \in X$. This ring provides the important link between the algebraic level (resp. sublevel) and the topological level (resp. colevel):

$$S'(X) \le \sigma(A_X) \le S(A_X) = S(X)$$

for any topological space with involution ε. Using this fact and some topological properties of the real spheres and the real Stiefel manifolds, Dai, Lam and Peng obtained their results mentioned above.

10. Hermitian forms over quaternions

In 1979-1980, Lu Hongwen published two papers ([43] and [44]) concerning analytic theory of Hermitian forms over quaternion, which actually were completed in 1965. He proved the following results:

1) The group of Hermitian unimodular matrices of rank n is generated by

$$V = \begin{pmatrix} 1 & i & \\ 0 & 1 & \\ & & I_{n-2} \end{pmatrix}, \quad S = \begin{pmatrix} h & 0 \\ 0 & I_{n-1} \end{pmatrix}, \quad T = \begin{pmatrix} 0 & I_{n-1} \\ 1 & 0 \end{pmatrix}$$

where I_m is the identity matrix of rank m, $h = (1-i-j-k)/2$, $i^2 = j^2 = -1$ and $ij = -ji = k$.

2) An arithmetic theory of Hermitian forms over quaternion is set up and Hasse-Minkowski principle is proved.

3) The fundamental theorem for Hermitian forms is proved, i.e., the mass formula holds, which coincides with the assertion that the Tamagawa number for Hermitian unitary group is 1.

References

[1] Arf, Cahit, Untersuchungen über quadratische Formen in Köpern der Characteuistik 2, I and II, J. Reine Angew. Math., 183, 148-167 (1941, and Rev. Fac. Sci. Univ. Istanbul (A) 8, 297-327 (1943).

[2] Chaundy, T.W., The arithmetic minima of positive quadratic forms, I, Quart. J. Math., (Oxford) Ser. 17 (1946), 166-192.

[3] Chu, Futsu and Ko, Chao, On the equivalence of positive definite quadratic and hermitian forms, Sci. Record, V. 2, No. 2 (1948), 148-155.

[4] Dai, Zongduo, On transitivity of subspaces in orthogonal geometry over fields of characteristic 2, Acta Math. Sinica 16 (1966), 545-560.

[5] Dai, Zongduo, Quadratic forms over fields of characteristic 2, Acta Math. Sinica 24 (1981), 383-389.

[6] Dai, Z.D., Lam, T.Y, and Peng, C.K., Levels in algebra and topology, Bulletin (New series) of AMS, V. 3 (1980), 845-848.

[7] Dai, Z.D., Lam, T.Y., et al., The Pythagoras numbers of some affine algebras and local algebras, J. Reine Angew. Math. 336 (1982), 45-82.

[8] Dai, Z.D., Lam, T.Y., Levels in algebra and topology, Comment. Math. Helv. 59 (1984), 376-424.

[9] Delone, B.N., Geometry of positive quadratic forms addendum, Uspehi Mat. Nauk 4 (1938).

[10] Erdös, P. and Ko, Chao, On definite forms which are not the sum of two definite or semi-definite forms, Acta Arith. 3 (1938), 102-122.

[11] Erdös, P. and Ko, Chao, Some results on definite quadratic forms, J. London Math. Soc., 13 (1938), 217-224.

[12] Hsia, J.S., Regular positive ternary quadratic forms, Mathematika 28 (1981), 231-238.

[13] Knebusch, M., Some open problems, in Conf. on quadratic forms - 1976, ed. by G. Orzech, Queen's Univ., 1977, p. 367.

[14] Ko, Chao, On a Waring's problem with squares of linear forms, Proc. London Math. Soc. 42 (1936), 171-185.

[15] Ko, Chao, On the representation of a quadratic form as a sum of squares of linear forms, Quart. J. of Math. (Oxford) 8 (1937), 81-98.

[16] Ko, Chao, Note on the representations of a quadratic form as a sum of squares of linear forms, Quart. J. of Math. (Oxford) 9 (1938), 32-33.

[17] Ko, Chao, On the positive definite quadratic forms with determinant unity, Acta Arith. 3 (1938), 79-85.

[18] Ko, Chao, On the decomposition of quadratic forms in six variables, Acta Arith. 3 (1938), 64-78.

[19] Ko, Chao, Determination of the class number of positive quadratic forms in nine variables with determinant unity, J. London Math. Soc. 13 (1938), 102-110.

[20] Ko, Chao, On the Meyer's theorem and the decomposition of quadratic forms, J. Chinese Math. Soc. 2 (1940), 209-224.

[21] Ko, Chao, On the decomposition of quadratic forms in eight variables, Sci. Record, V. 1, No. 1-2 (1942), 33-36.

[22] Ko, Chao, On the decomposition of quadratic forms in seven variables, Sci. Record, V. 1, No. 1-2 (1942), 30-33.

[23] Ko, Chao, On the decomposition of quadratic forms in eight variables, The Sci. Report of Nat. Tsing Hua Univ. Series A, 4 (1947), 337-340.

[24] Ko, Chao, Decomposition of ternary quadratic forms as sum of squares, Acta Sci. of Sichuan Univ., V. 56, No. 1 (1956), 17-30.

[25] Ko, Chao, A problem on quadratic forms, Acta Sci. of Sichuan Univ., V. 57, No. 1 (1957), 23-31.

[26] Ko, Chao, Definite quadratic forms of 12 and 13 variables with determinant unity, Acta Sci. of Sichuan Univ., V. 58, No. 1 (1958), 15-32.

[27] Ko, Chao, Definite quadratic forms of 14 variables with determinant unity, Acta Sci. of Sichuan Univ., V. 59, No. 1 (1959), 31-58.

[28] Ko, Chao, Definite quadratic forms of 15 variables with determinant unity, Acta Sci. of Sichuan Univ., V. 59, No. 2 (1959), 23-38.

[29] Ko, Chao, Positive definite unimodular quadratic forms, Acta Sci. of Sichuan Univ., V. 59, No. 3 (1959), 35-42.

[30] Ko, Chao, et al., Positive definite quadratic forms in 6 variables with determinant 25, Acta Sci. of Sichuan Univ., V. 64, No. 2 (1964), 11-26.

[31] Ko, Chao and Wang, S.C., Table of primitive positive quaternary quadratic forms with determinant ≤ 25, Sci. Record, V. 1, No. 1-2 (1942), 54-58.

[32] Ko, Chao, Xuan, Huawu and Zheng, Du, Positive definite quinary quadratic forms of determinants ≤ 50, Acta Sci. of Sichuan Univ., V. 60, No. 1 (1960), 1-25.

[33] Ko, Chao and Zheng, Du, Positive definite primitive quaternary quadratic forms of determinants ≤ 50, Acta Sci. of Sichuan Univ., V. 59, No. 5 (1959), 1-14.

[34] Ko, Chao and Zheng, Dexun, On minimum of positive definite quadratic forms, Acta Sci. of Sichuan Univ., V. 63, No. 2 (1963), 43-62.

[35] Ko, Chao and Zheng, Dexun, Positive definite quadratic forms representing the same numbers, Acta Sci. of Sichuan Univ., V. 89, No. 2-3 (1978), 5-14.

[36] Ko, Chao and Zheng, Dexun, On positive definite ternary quadratic forms representing the same numbers, Acta Sci. of Sichuan Univ., V. 80, No. 4 (1980), 1-8.

[37] Ko, Chao and Zheng, Dexun, On quadratic forms of type $a(x^2+y^2)+bz^2$ representing the same numbers, Acta Sci. of Sichuan Univ., V. 83, No. 3 (1983), 1-8.

[38] Kula, M., Fields with nontrivial Kaplansky's radical and finite square class number, Acta Arith. 38 (1980/81), 411-418.

[39] Kula, M., Szczepanik, L., Szymiczek, K., Quadratic forms over formally real fields with eight square classes, Manuscripta Math. 39 (1979), 295-303.

[40] Li, Delang, A family of fields with finite square class number, Scientia Sinica, 13 (1980), 1226-1236.

[41] Li, Delang, Indefinite binary forms representing the same numbers, Math. Proc. Camb. Phil. Soc., 92 (1982), 29-33.

[42] Li, Delang, Representation of numbers by binary quadratic forms, Acta Math. Sinica, New series, 3, (1987), 58-65.

[43] Lu, Hongwen, The analytic theory of Hermitian forms (I), J. China Univ. Sci. Tech., 9 (1979), 66-77.

[44] Lu, Hongwen, The analytic theory of Hermitian forms (II) - (IV), J. China Univ. Sci. Tech., 10 (1980), 1-30.

[45] Mordell, An application of quaternions to the representation of a binary quadratic form as a sum of four linear squares, Quart. J. Math. Oxford Ser. 8 (1937), 58-61.

[46] O'Meara, O.T., The construction of indecomposable positive definite quadratic forms, J. Reine Angew. Math. 276 (1975), 99-123.

[47] Szymiczek, K., Quadratic forms over fields with finite square class number, Acta Arith. 28 (1976), 195-221.

[48] Timofeev, V.N., On positive quadratic forms representing the same numbers, (Russian) Uspehi Mat. Nauk 18 (1963) 4 (12), 191-193.

[49] Watson, G.L., Integral quadratic forms, Cambridge Univ. Press. (1960), p. 115.

[50] Watson, G.L., Determination of a binary quadratic form by its values at integer points, Mathematika 26 (1979), no. 1, 72-75.

[51] Witt, E., Theorie der quadratischen Formen in beliebigen Körpern, J. Reine Angew. Math. 176 (1937), 31-44.

[52] Zheng, Dexun, On quadratic forms of the type $a(x^2+2y^2)+bz^2$ representing the same numbers, Acta Sci. of Sichuan Univ., V. 85, No. 3 (1985), 25-29.

[53] Zhu, Fuzu, On the construction of non-decomposable positive definite unimodular quadratic forms, Scientia Sinica (Series A), 1986, 690-700.

[54] Zhu, Fuzu, On the non-decomposability and indecomposability, Scientia Sinica (Series A), 1986, 1037-1043.

Mathematics Department
Sichuan University
Chengdu, Sichuan
China

Mathematics Department
University of Science and Technology of China
Hefei, Anhwei
China

Contemporary Mathematics
Volume **77**, 1988

SMALL PRIME SOLUTIONS OF LINEAR EQUATIONS AND THE EXCEPTIONAL SET IN GOLDBACH'S PROBLEM

Liu Mingchit and Tsang Kaiman

Abstract

Let a_1, a_2, a_3 be any three non-zero integers such that

$$\gcd(a_1, a_2, a_3) = 1. \tag{1}$$

Let b be an integer satisfying

$$b \equiv a_1 + a_2 + a_3 \pmod 2 \quad \text{and} \quad \gcd(b, a_i, a_j) = 1 \quad \text{for any} \quad 1 \le i < j \le 3. \tag{2}$$

Write $B := \max(3, |a_1|, |a_2|, |a_3|)$. We can prove:

Theorem 1. Suppose a_1, a_2, a_3 are positive and satisfy (1). Then for any small positive ε, there exists an effective positive constant $A = A(\varepsilon)$ such that, whenever b satisfies (2) and $b \ge B^A$, the equation

$$a_1p_1 + a_2p_2 + a_3p_3 = b \tag{3}$$

has at least $b^{2-\varepsilon}$ solutions in primes p_1, p_2 and p_3.

Theorem 2. Suppose a_1, a_2, a_3 are not all of the same sign and satisfy (1). Then for any small positive ε, there exists an effective positive constant $A = A(\varepsilon)$ such that, if b satisfies (2) and $N \ge 3|b| + B^A$ equation (3) has at least $N^{2-\varepsilon}$ solutions in primes p_1, p_2, p_3 not exceeding N. In particular, there are solutions p_1, p_2, p_3 of (3) satisfying

$$\max(p_1, p_2, p_3) \le 3|b| + B^A. \tag{4}$$

Theorem 1 is a generalization of Vinogradov's famous three primes theorem. The main interest lies in the lower bound $b \ge B^A$ for which (3) is solvable. Apart from the value of A, this is sharp because a trivial lower bound for b is $2B$.

The problem on (4), the bounds for prime solutions of those equations

0271-4132/88 $1.00 + $.25 per page

mentioned in Theorem 2, was first studied by Baker [1] in connection with some diophantine inequalities. Later developments show that this problem has independent interest. In [1], for $b = 1$ or 2, an upper bound of the form C^{B^δ} for p_j was obtained where δ is any positive number and C is a positive constant depending on δ only. Instead of (1), assuming the stronger condition

$$\gcd(a_i, a_j) = 1 \quad \text{for} \quad 1 \le i < j \le 3, \tag{5}$$

the first author [4] could replace this bound by B^K for some noneffective constant K. The proof in [4] combined the circle method with a linear sieve. Although this combination was successful in giving the better bound B^K, the stronger assumption (5) as well as the noneffectiveness of K seem unavoidable in his argument. Our Theorem 2 removes such unnecessary restrictions on the a_j's and the blemish on K. Furthermore, our integer b satisfies (2) and is otherwise unrestricted. The example with $a_1 = b = 1$, $a_2 = -1$ and $a_3 = -B$ (for any odd integer $B \ge 3$) shows that $\max(p_1, p_2, p_3) \ge p_1 = 1 + p_2 + Bp_3 \ge 3 + 2B$. Thus, apart from the exact value of A, the estimate (4) is sharp. Baker's original problem is thus completely solved and extended.

Theorems 1 and 2 are proved by a refinement of the circle method. The main difficulty lies in the treatment of the various dominating terms under the influence of the complex zeros of L-functions. This requires very detailed asymptotic analysis on various terms, and the dominating terms have to be arranged in a special way. Moreover, we need the very sharp estimate $\sum_{q \le T} \sum^*_{\chi(\mathrm{mod}\ q)} N_\chi(\sigma, T) \ll T^{c(1-\sigma)}$ of Gallagher [3] concerning the distribution of the zeros of $L(S, \chi)$ near the line $\sigma = 1$. In essence, it replaces the role of the Generalized Riemann Hypothesis.

Our method can be extended to solving a system of linear equations in small primes. As we are aware, so far nothing has been known on bounds for prime solutions of simultaneous linear equations and previous noteworthy work on the solubility of these equations in primes can only be found in [2,6].

Consider a pair of linear equations in 5 prime variables:

$$\sum_{j=1}^{5} a_{\lambda j} p_j = b_\lambda; \quad \lambda = 1, 2, \tag{6}$$

where $a_{\lambda j}$ and b_λ are integers. We put $B_1 := \max_{\substack{\lambda=1,2 \\ 1 \le j \le 5}} \{100, |a_{\lambda j}|\}$ and for each prime p and integer $s = 3$ or 5 define

$$\mathcal{N}_s(p) := \mathrm{card}\{(\ell_1,\dots,\ell_s) : 1 \le \ell_j \le p-1;\ \sum_{j=1}^{s} a_{\lambda j}\ell_j \equiv b_\lambda \pmod p;\ \lambda = 1,2\}.$$

For $1 \le i,j$ let

$$\Delta_{ij} = \begin{vmatrix} a_{1i} & a_{1j} \\ a_{2i} & a_{2j} \end{vmatrix} \quad \text{and} \quad \Delta_{ib} = \begin{vmatrix} a_{1i} & b_1 \\ a_{2i} & b_2 \end{vmatrix}.$$

We can prove

Theorem 3. Suppose the coefficients $a_{\lambda j}, b_\lambda$ $(\lambda = 1,2;\ j = 1,\dots,5)$ satisfy the following hypotheses:

(i) $\Delta_{ij} \ne 0$ for $1 \le i < j \le 5$,

(ii) $\gcd(\Delta_{ij})_{1\le i<j\le 5} = 1$,

(iii) $\mathcal{N}_5(p) > 0$ for each prime p,

(iv) there exist positive real numbers $y_1,\dots,y_5$ satisfying

$$\sum_{j=1}^{5} a_{\lambda j} y_j = 0, \quad \text{for } \lambda = 1,2.$$

Then given any small positive ε, there exists an effective positive constant $A = A(\varepsilon)$ such that, for any

$$N \ge (|b_1| + |b_2| + 1)B_1^A \tag{7}$$

the pair of equations in (6) has at least $N^{3-\varepsilon}$ solutions in primes $p_1,\dots,p_5$ with $\max_{1\le j\le 5}\{p_j\} \le N$. In particular, there are solutions $p_1,\dots,p_5$ of (6) satisfying

$$\max_{1\le j\le 5}\{p_j\} \le (|b_1| + |b_2| + 1)B_1^A. \tag{8}$$

The hypotheses (i) and (ii) are natural extensions of (1), while (iii) is equivalent to the positiveness of the singular series. This corresponds to the hypothesis in (2) and we can prove a corresponding criterion for which (iii) holds, e.g., for $p \ge 5$, $N_5(p) = 0$ iff there exists $1 \le i \le 5$ such that, for all $k,j \in \{1,\dots,5,b\}\setminus\{i\}$, we have $\Delta_{kj} \equiv 0 \pmod p$. The assumption in (iv) is needed to ensure the positiveness of an integral occurred in the main term. In fact, we can show that (iv) holds iff there exist $i,j,k \in \{1,2,\dots,5\}$ such that all $\Delta_{ij}, \Delta_{jk}, \Delta_{ki}$ are positive.

The significance of Theorem 3 lies in the lower limit (7) for N and the upper bound of the solutions in (8). Apart from the exact value of A, these estimates are sharp, as shown by the following example: for any positive $B \equiv 4 \pmod 6$, the pair of equations

$$Bp_1 + p_2 + p_3 + 2p_4 - 3p_5 = 1, \quad Bp_1 - p_2 + 3p_3 + p_4 + 2p_5 = 1$$

satisfies the hypotheses (i) - (iv) in Theorem 3. Now clearly, $\max_{1\le j\le 5}\{p_j\} \ge p_2 = Bp_1 + 3p_3 + p_4 + 2p_5 - 1 \ge 2B + 11$.

If we let $E(X)$ denote the set of positive even integers not exceeding X which cannot be written as a sum of two primes, then Goldbach's conjecture asserts that $E(X) = \{2\}$ for all $X \ge 2$. The difficult nature of this famous conjecture warrants the search of good upper bounds for $\operatorname{card} E(X)$ a worthwhile investigation. This indeed has attracted much attention in the field for the past fifty years. Hitherto the best result was obtained by Montgomery and Vaughan [5]. They proved that for some positive absolute constant δ, $\operatorname{card} E(X) \le X^{1-\delta}$ holds for all large X. The first author [4] then extended their result to the general case of representing integers in the form $a_1p_1 + a_2p_2$ where a_1, a_2 are given relatively prime positive integers (see the Corollary below). In this paper, we consider the general question of representing simultaneously a pair of integers (b_1, b_2) by two linear forms in three prime variables:

$$\sum_{j=1}^{3} a_{\lambda j}p_j = b_\lambda; \quad \lambda = 1,2.$$

By injecting further ideas into the method we have employed in the proof of Theorems 3, we can prove

Theorem 4. Let the coefficients $a_{\lambda j}$'s satisfy the conditions:

(i) $\Delta_{12}\Delta_{13}\Delta_{23} \ne 0$, $\gcd(\Delta_{12}, \Delta_{13}, \Delta_{23}) = 1$,

(ii) there exist positive real numbers y_1, y_2, y_3 such that

$$\sum_{j=1}^{3} a_{\lambda j}y_j > 0, \quad \text{for } \lambda = 1,2.$$

Let $B_2 := \max_{\substack{\lambda=1,2\\ j=1,2,3}} \{3|a_{\lambda j}|\}$ and for any large X, let

$$\mathcal{W}(X) := \{(b_1,b_2) : 0 \le b_1,b_2 \le X;\ \mathcal{N}_3(p) > 0 \text{ for all primes } p \text{ and the system } \sum_{j=1}^{3} a_{\lambda j}y_j = b_\lambda;\ \lambda = 1,2 \text{ is solvable in positive real } y_1,y_2,y_3\}.$$

Then (I) we have

$$\operatorname{card} \mathcal{W}(X) \gg X^2(B_2^2 \log\log B_2)^{-2};$$

(II) there exist effective positive absolute constants A and δ such that, for any $X \ge B_2^A$, the set

$$E_1(X) := \{(b_1,b_2) \in \mathcal{W}(X) : \sum_{j=1}^{3} a_{\lambda j}p_j = b_\lambda;\ \lambda = 1,2 \text{ is not solvable}\}.$$

satisfies

$$\operatorname{card} E_1(X) \le X^{2-\delta}.$$

Applying our Theorem 4 we can obtain the results mentioned above on the exceptional set $E(X)$. That is:

Corollary (essentially in [4,5]). Let a_1,a_2 be non-zero integers, not both negative, and $\gcd(a_1,a_2) = 1$. Then there exist positive effective absolute constants δ and A such that for any $X \ge \max(3|a_1|,3|a_2|)^A$, the set

$$E(X) := \{0 \le b \le X : \gcd(b,a_1a_2) = 1,\ b \equiv a_1 + a_2 \pmod 2 \text{ and } b = a_1p_1 + a_2p_2 \text{ has no prime solutions } p_1 \text{ and } p_2\}$$

satisfies

$$\operatorname{card} E(X) \le X^{1-\delta}.$$

This shows that our Theorem 4 is indeed a stronger result than those obtained in [4,5] on the exceptional set of Goldbach's problem.

We notice that results in Theorems 3 and 4 can be extended to $n\ (\ge 3)$ linear equations in $2n+1$ and $n+1$ variables respectively.

REFERENCES

1. Baker, A.: On some diophantine inequalities involving primes, J. reine angew. Math. 228, 166-181 (1967).

2. Van der Corput, J.G.: Propriétés additives I, Acta Arith. 3, 180-234 (1939).

3. Gallagher, P.X.: A large sieve density estimate near $\sigma = 1$, Invent. Math. 11, 329-339 (1970).

4. Liu, M.C.: An improved bound for prime solutions of some ternary equations, Math. Z. 194, 573-583 (1987).

5. Montgomery, H.L., Vaughan, R.C.: The exceptional set in Goldbach's problem, Acta Arith. 27, 353-370 (1975).

6. Wu, Fang: On the solutions of the systems of linear equations with prime variables, Acta Math. Sinica 7, 102-122 (1957) (in Chinese with English summary).

Department of Mathematics
University of Hong Kong
Pokfulam Road, Hong Kong

Contemporary Mathematics
Volume 77, 1988

ON THE RELATIVE TRACE FORMULA

K.F. Lai

Arthur ([1]) generalized the Selberg trace formula to an arbitrary reductive group G. His trace formula is an equality of distributions, with the class (or called $\mathfrak{o}$) expansions on one side and the representations (or called χ) expansions on the other. In Jacquet & Lai we give another generalization of the Selberg trace formula for GL(2), namely instead of integrating the kernel of the right regular representation K(x,y) over the group G embedded diagonally in G×G, i.e., the usual integration of K(x,x) over G, we integrate over an off diagonal subgroup H of G×G instead of G. However, in Jacquet and Lai as in the classical case of SL(2) we attack the convergence problem of integration directly without the sophisticated means of Arthur. In this paper we propose an analogue of Arthur [1] for the class expansion side of a relative trace formula of an arbitrary reductive group. We hope that this would provide an alternative route to study the base change problem proposed by Langlands, in particular to obtain results for compact Shimura varieties where there are no Eisenstein series. For example this technique allows one to prove Tate's conjecture on algebraic cycles for a family of compact Shimura surfaces (Lai [3]).

Let F be an algebraic number field and E be a finite Galois extension of F. We write $\mathbb{A}$ for the adeles of F and $\mathbb{A}_E$ for the adeles of E. To simplify notation we use bold face **G** to denote an algebraic group defined over F and write G for its group of F rational points, $G_{\mathbb{A}}$ for its group of F adelic points.

Let **G** be a connected reductive algebraic group defined over F. Consider **G** as an algebraic group over E and apply to it the Weil restriction functor from E to F to obtain $\tilde{G}_{\mathbb{A}}$. For each rational character χ of $\tilde{G}_{\mathbb{A}}$ defined over F, we define the homomorphism $|\chi|$ by

$$|\chi|(x) = \prod_{\mathfrak{v}} |\chi(x_{\mathfrak{v}})|_{\mathfrak{v}}, \quad x = \prod x_{\mathfrak{v}} \in \tilde{G}_{\mathbb{A}}.$$

The intersection of the kernel of all the $|\chi|$ is denoted by $\tilde{G}_{\mathbb{A}}$. The locally compact group $\tilde{G}_{\mathbb{A}}$ contains $\tilde{G}$ as a discrete subgroup. Let R be the right regular representation of $\tilde{G}_{\mathbb{A}}$ in the Hilbert space $L^2(\tilde{G}\backslash\tilde{G}_{\mathbb{A}})$, and f be a smooth compactly supported function on $\tilde{G}_{\mathbb{A}}$. Then for $\phi \in L^2(\tilde{G}\backslash\tilde{G}_{\mathbb{A}})$ the convolution operator $R(f)$ is given by

$$R(f)\phi(x) = \int_{\tilde{G}_{\mathbb{A}}} f(y)R(y)\phi(x)dy$$

$$= \int_{\tilde{G}\backslash\tilde{G}_{\mathbb{A}}} K(x,y)\phi(y)dy,$$

where

$$K(x,y) = \sum_{\gamma\in\tilde{G}} f(x^{-1}\gamma y)$$

is the kernel of the integral operator $R(f)$. We expand the kernel K in two ways, namely according to the semisimple conjugacy classes o in $\tilde{G}$ and the equivalence classes χ of cuspidal automorphic representations of Levi components of parabolic subgroups of $\tilde{G}$. We obtain an identity

$$\sum K_o(x,y) = K(x,y) = \sum K_\chi(x,y).$$

The problem of trace formula is to integrate both sides of the above identity (with $x = y$) over $\tilde{G}\backslash\tilde{G}_{\mathbb{A}}$ which is not necessarily compact and so the integrals may diverge. Imitating the truncation operator introduced by Langlands for Eisenstein series, Arthur ([1]) introduced for the first time the correct formula for the modification of the above identity and the modified expansions of the kernel is integrable.

We can consider G as a subgroup of $\tilde{G}$ and we would like to study the relative trace:

$$\int_{(G\backslash G_{\mathbb{A}})^2} K(x,y)dxdy.$$

Here the kernel K is the one given above for the group $\tilde{G}$. In this paper we introduce an analogue of Arthur's modification of the kernel K and carry out the rest of the program of the relative trace formula. In particular if T is a sufficiently positive element in the Lie algebra of the split component of the minimal parabolic and we write λ^T for the relative truncation we can prove the following theorem.

Theorem. In the notations given above we have

$$\sum_{\mathfrak{o}\in\mathcal{O}}\int_{(G\backslash G_{\mathbb{A}})^2}\lambda^T K_{\mathfrak{o}}(x,y)dxdy = \sum_{\chi\in\mathfrak{X}}\int_{(G\backslash G_{\mathbb{A}})^2}\lambda^T K_{0}(x,y)dxdy.$$

References

[1] J. Arthur, A trace formula for reductive groups I, Duke Math. Journal, 45 (1978), 911-952.

[2] H. Jacquet and K.F. Lai, A relative trace formula, Compositio Math. 54 (1985), 243-301.

[3] K.F. Lai, Algebraic cycles on compact Shimura surfaces, Math. Zeit., 189 (1985), 593-602.

Mathematics Department
Chinese University
Hong Kong

Contemporary Mathematics
Volume 77, 1988

KLOOSTERMAN INTEGRALS AND BASE CHANGE*

Ye Yangbo

1. The purpose of this note is to announce the results in the thesis of the author [5].

Let F be a finite algebraic number field and $E = F(\sqrt{\tau})$ a quadratic extension of F with $\tau \in F^{x}$. We consider the representations of $GL(2,E_{\mathbb{A}})$ and $GL(2,F_{\mathbb{A}})$. One can prove the following theorem by the theory of base change and Asai L-functions.

Theorem. An automorphic irreducible cuspidal representation π of $Z_{E_{\mathbb{A}}}\backslash GL(2,E_{\mathbb{A}})$ is said to be distinguished if there is a smooth function f' in the space of π such that

$$\int_{Z_{F_{\mathbb{A}}} GL(2,F)\backslash GL(2,F_{\mathbb{A}})} f'(x)dx \neq 0.$$

Then π is the base change of an automorphic cuspidal irreducible representation π of $GL(2,F_{\mathbb{A}})$ whose central character is the quadratic idele class character η of F attached to E if and only if π is distinguished.

We will given a different proof of part of that theorem by establishing a relative trace formual. Indeed, we will assume that every infinite place of F splits in E and show that π is the lifting if it is distinguished.

2. Let $f' = \pi f'_w$ be a smooth function on $GL(2,E_{\mathbb{A}})$, trivial on the center $Z_{E_{\mathbb{A}}}$, of compact support in $Z_{E_{\mathbb{A}}}\backslash GL(2,E_{\mathbb{A}})$. We assume that f'_w is the characteristic function of $Z_{E_w} GL(2,R_w)$ for almost every place w of E. Similarly, set $f = \pi f_v$ to be a smooth function on $GL(2,F_{\mathbb{A}})$, with central character η, of compact support modulo $Z_{F_{\mathbb{A}}}$. Suppose, for almost every

* A complete proof of the theorem has been furnished by the author using our relative trace formula. The limit of the truncation in the proposition is actually equal to the integration of the Eisenstein kernel function.

place v, that f_v equals 1 on $GL(2,R_v)$ and vanishes outside $Z_{F_v} GL(2,R_v)$.

We denote by $K^{f'}_{cusp}(g,h)$ the kernel of the representation of $GL(2,E_{\mathbb{A}})$ associated with f', on the space of cuspidal functions φ on $GL(2,E_{\mathbb{A}})$ such that

$$\varphi(\nu zg) = \varphi(g)$$

with $\nu \in GL(2,E)$, $z \in Z_{E_{\mathbb{A}}}$ and $g \in GL(2,E_{\mathbb{A}})$, and that

$$\int_{Z_{E_{\mathbb{A}}} GL(2,E)\backslash GL(2,E_{\mathbb{A}})} |\varphi(g)|^2 dg < +\infty.$$

Likewise $K^{f}_{cusp}(g,h)$ denotes the kernel of the representation of $GL(2,F_{\mathbb{A}})$ associated with f on the space of cuspidal functions φ on $GL(2,F_{\mathbb{A}})$ such that

$$\varphi(\nu zg) = \eta(z)\varphi(g)$$

with $\nu \in GL(2,F)$, $z \in Z_{F_{\mathbb{A}}}$ and $g \in GL(2,F_{\mathbb{A}})$, and that

$$\int_{Z_{F_{\mathbb{A}}} GL(2,F)\backslash GL(2,F_{\mathbb{A}})} |\varphi(g)|^2 dg < +\infty.$$

3. For a function h on $Z_{E_{\mathbb{A}}} GL(2,E)\backslash GL(2,E_{\mathbb{A}})$ and a constant $c > 1$, we define the relative truncation $T^c h$ as a function on $Z_{F_{\mathbb{A}}} GL(2,F)\backslash GL(2,F_{\mathbb{A}})$ by

$$T^c h(g) = h(g) - \sum_{\nu \in p_F\backslash GL(2,F)} \chi_c(H(\nu g)) \int_{E\backslash E_{\mathbb{A}}} h\left(\begin{pmatrix} 1 & x \\ 0 & 1 \end{pmatrix} \nu g\right) dx$$

where χ_c is the characteristic function of $[c,\infty)$ and

$$H\left(\begin{pmatrix} a & x \\ 0 & b \end{pmatrix} k\right) = \left|\frac{a}{b}\right|_{E_{\mathbb{A}}}.$$

<u>Proposition</u>. Given f' above, we can find a function f with the properties described previously such that the following holds:

$$\lim_{c\to+\infty} \int_{Z_{F_{\mathbb{A}}} GL(2,F)\backslash GL(2,F_{\mathbb{A}})} \int_{E\backslash E_{\mathbb{A}}} T^c_g \sum_{\xi \in Z_E\backslash GL(2,E)} f'\left(g^{-1}\xi \begin{pmatrix} 1 & x \\ 0 & 1 \end{pmatrix}\right)\varphi'(x) dg dx =$$

$$= c_F \int_{F\backslash F_{\mathbb{A}}} \int_{F\backslash F_{\mathbb{A}}} \sum_{\xi\in Z_F\backslash GL(2,F)} f\left(\begin{pmatrix}1 & -x\\0 & 1\end{pmatrix}\xi\begin{pmatrix}1 & y\\0 & 1\end{pmatrix}\right)\psi(y-x)dxdy \qquad (1)$$

where ψ is a character of $F_{\mathbb{A}}$, $\psi' = \psi\circ tr$ and c_F is a non-zero constant.

By Gelbart and Jacquet [2], the cuspidal kernels can be expressed as

$$K^{f'}_{cusp}(g,h) = \sum_{\xi\in Z_E\backslash GL(2,E)} f'(g^{-1}\xi h) - K^{f'}_{eis}(g,h) - K^{f'}_{sp}(g,h)$$

and

$$K^{f}_{cusp}(g,h) = \sum_{\xi\in Z_F\backslash GL(2,F)} f(g^{-1}\xi h) - K^{f}_{eis}(g,h) - K^{f}_{sp}(g,h).$$

Since

$$T^c_g K^{f'}_{cusp}(g,h) = K^{f'}_{cusp}(g,h)$$

$$\iint T^c_g K^{f'}_{sp}\left(g, \begin{pmatrix}1 & x\\0 & 1\end{pmatrix}\right)\psi'(x)dgdx = 0$$

and

$$\iint K^{f}_{sp}\left(\begin{pmatrix}1 & x\\0 & 1\end{pmatrix}, \begin{pmatrix}1 & y\\0 & 1\end{pmatrix}\right)\psi(y-x)dxdy = 0$$

we get

<u>Corollary</u>. For the functions f' and f as in the proposition, we have

$$\iint K^{f'}_{cusp}\left(g, \begin{pmatrix}1 & x\\0 & 1\end{pmatrix}\right)\psi'(x)dgdx - c_F\iint K^{f}_{cusp}\left(\begin{pmatrix}1 & x\\0 & 1\end{pmatrix}, \begin{pmatrix}1 & y\\0 & 1\end{pmatrix}\right)\psi(y-x)dxdy =$$

$$(2)$$

$$= -\lim_{c\to\infty} \iint T^c_g K^{f'}_{eis}\left(g, \begin{pmatrix}1 & x\\0 & 1\end{pmatrix}\right)\psi'(x)dgdx + c_F\iint K^{f}_{eis}\left(\begin{pmatrix}1 & x\\0 & 1\end{pmatrix}, \begin{pmatrix}1 & y\\0 & 1\end{pmatrix}\right)\psi(y-x)dxdy.$$

4. By methods similar to those in Jacquet and Lai [3], we can calculate the integrals of $T^c K^{f'}_{eis}$ and K^{f}_{eis}.

Let us denote by S a finite set of places of F containing all the infinite and ramified places and write

$$f^S = \prod_{v\notin S} f_v,\quad f_S = \prod_{v\in S} f_v,\quad F_S = \prod_{v\in S} F_v, \quad \text{etc.}$$

Let T be the set of places of E lying above S. Then

$$c_F\iint K^f_{eis}\left(\begin{pmatrix}1 & x\\ 0 & 1\end{pmatrix},\begin{pmatrix}1 & y\\ 0 & 1\end{pmatrix}\right)\psi(y-x)dxdy = \sum_{\mu\nu=\eta}\int_{-\infty}^{\infty}\pi(it,\mu,\nu)^{\wedge}(f^S)\cdot a(it,\mu,\nu)dt$$

where $a(it,\mu,\nu)$ is a constant coefficient. Note that here $\pi(it,\mu,\nu)^{\wedge}(f^S)$ is the unitary homomorphism corresponding to the representation $\pi(it,\mu,\nu)$, from the Hecke algebra $\mathcal{H}(GL(2,F^S,K^S)$ of compactly supported functions on $GL(2,F^S)$, bi-invariant under K^S, into $\mathbb{C}$. Similarly,

$$-\lim_{c\to\infty}\iint T^c_g K^{f'}_{eis}\left(g,\begin{pmatrix}1 & x\\ 0 & 1\end{pmatrix}\right)\psi'(x)dgdx$$

$$=\sum_{\mu}\int_{-\infty}^{\infty}\pi(it,\mu,\mu^{-1})^{\wedge}(f'^T)b(it,\mu)dt+\sum_{\mu}\pi(0,\mu,\mu^{-1})^{\wedge}(f'^T)c(\mu).$$

We can choose χ such that

$$\pi(it,\mu,\nu)^{\wedge}(f^S) = \pi(it,\chi,\chi^{-1})^{\wedge}(f'^T). \tag{3}$$

Hence the right hand side of (2) is actually

$$\sum_{\mu}\int_{-\infty}^{\infty}\pi(it,\mu,\mu^{-1})^{\wedge}(f'^T)d(it,\mu)dt-\sum_{\mu}\pi(0,\mu,\mu^{-1})^{\wedge}(f'^T)c(\mu).$$

5. As in Jacquet and Lai [3], if σ' is a cuspidal irreducible representation of $Z_{E_{\mathbb{A}}}\backslash GL(2,E_{\mathbb{A}})$ containing the unit representation of K^T, we denote by $V(\sigma')$ the subspace with the orthonormal basis $\{\Phi_j\}$, of the space of σ' consisting of forms invariant under K^T and by σ'_T the corresponding representation of $Z_{E_T}\backslash GL(2,E_T)$ on $V(\sigma')$. Then

$$K^{f'}_{cusp}(g,h) = \sum_{\sigma'}\sigma'^{\wedge}(f'^T)\sum_j\sigma'_T(f'_T)\Phi_j(g)\bar{\Phi}_j(h)$$

where $\sigma'^{\wedge}(f'^T)$ is the unitary homomorphism as in (3).

By the same reason, if p goes over all the cuspidal irreducible representations of $GL(2,F_{\mathbb{A}})$, with central character η, containing the unit representation of K^S, then

$$K^f_{cusp}(g,h) = \sum_p p^{\wedge}(f^S)\sum_j p_S(f_S)\phi_j(g)\bar{\varphi}_j(h)$$

where $\{\varphi_j\}$ is an orthonormal basis of $V(p)$. Similar to (3), we have

$$p^{\wedge}(f^S) = p'^{\wedge}(f'^T). \tag{4}$$

If p' is cuspidal, we can show that p' is the lifting of p. Therefore the left hand side of (2) equals

$$\sum_{\sigma'} \sigma'(f'^T)m(\sigma') - \sum_{p'} p'^{\wedge}(f'^T)n(p') - \sum_{p''} p''^{\wedge}(f'^T)e(p'')$$

where $m(\sigma'), n(p')$ and $e(p'')$ are constant coefficients. The sum with respect to σ' is taken over all the cuspidal irreducible representations of $Z_{E_{\mathbb{A}}}\backslash GL(2,E_{\mathbb{A}})$ which contain the unit representation of K^T, p' goes over all the cuspidal irreducible representations of $Z_{E_{\mathbb{A}}}\backslash GL(2,E_{\mathbb{A}})$ which are lifted from the cuspidal irreducible representations of $GL(2,F_{\mathbb{A}})$ with central character η and contain the unit representation of K^S, and p'' is over all the non-cuspidal irreducible representations of $Z_{E_{\mathbb{A}}}\backslash GL(2,E_{\mathbb{A}})$ obtained by (4).

Therefore from (2) we have

$$\sum_{\sigma'} \sigma'^{\wedge}(f'^T)m(\sigma') - \sum_{p'} p'^{\wedge}(f'^T)n(p') =$$

$$= \sum_{p''} p''^{\wedge}(f'^T)e(p'') + \sum_{\mu} \int_{-\infty}^{\infty} \pi(it,\mu,\mu^{-1})^{\wedge}(f'^T)d(it,\mu)dt \tag{5}$$

$$- \sum_{\mu} \pi(0,\mu,\mu^{-1})^{\wedge}(f'^T)c(\mu).$$

By Cartier [1], two characters are equal if and only if the corresponding representations are equivalent. Hence the characters from the two sides of (5) are not the same. Using the principle of linear independence of characters [4], we conclude that both sides of (5) must vanish, that is

$$\sum_{\sigma'} \sigma'^{\wedge}(f'^T)m(\sigma') = \sum_{p'} p'^{\wedge}(f'^T)n(p'). \tag{6}$$

6. Let π be an automorphic irreducible cuspidal representation of $Z_{E_{\mathbb{A}}}\backslash GL(2,E_{\mathbb{A}})$, containing the unit representation of $K^{T'}$. If π is distinguished, the restriction of B, where

$$B(\Phi) = \int_{Z_{F_{\mathbb{A}}} GL(2,F)\backslash GL(2,F_{\mathbb{A}})} \Phi(g)dg \tag{7}$$

to the space of π is a non-zero $GL(2,F_{\mathbb{A}})$-invariant form. Then there is a finite set T of places of E containing T' and all the infinite and ramified places which do not split in E, such that the restriction of B to V_T is non-zero, where V_T is the space of forms in the space of π which are invariant under K^T. We use this T to set up (6). Note that

$$m(\pi) = \int_{Z_{F_{\mathbb{A}}} GL(2,F)\backslash GL(2,F_{\mathbb{A}})} \pi_T(f'_T)\left[\sum_j \int_{E\backslash E_{\mathbb{A}}} \Phi_j\begin{pmatrix}1 & x\\ 0 & 1\end{pmatrix}\bar{\psi}'(x)dx\bar{\Phi}_j\right](g)dg.$$

Since $\pi_T(f'_T)$ acts on V_T transitively, we can choose f'_T such that $M(\pi) \neq 0$. Because of (6), there must be a representation p' from the right hand side such that

$$p' = \pi, \quad n(p') = m(\pi)$$

by the principle of linear independence of characters. Hence π is lifted from a representation with central character η.

7. The Proof of the Proposition.

Using the double coset decomposition

$$GL(2,E) = \bigcup_{\alpha\in E^{\times}/E^0} Z_E GL(2,F)\begin{pmatrix}\sqrt{\tau} & \alpha\\ 1 & 0\end{pmatrix}N_E \cup \bigcup_{e,f\in F} Z_E GL(2,F)\begin{pmatrix}1 & f\\ 0 & e+\sqrt{\tau}\end{pmatrix} \cup$$
$$\cup \bigcup_{e\in F} Z_E GL(2,F)\begin{pmatrix}1 & e\sqrt{\tau}\\ 0 & 1\end{pmatrix}$$

where $E^0 = \{\varepsilon \in E^{\times} | N_{E/F}(\varepsilon) = 1\}$ we find that

$$\lim_{c\to+\infty}\iint T_g^c \sum f'(g^{-1}\xi\begin{pmatrix}1 & x\\ 0 & 1\end{pmatrix})\psi'(x)dgdx =$$

$$= \sum_{\beta\in E^0/E^{\times}} \int_{Z_{F_{\mathbb{A}}}\backslash GL(2,F_{\mathbb{A}})}\int_{E_{\mathbb{A}}} f'(g^{-1}\begin{pmatrix}\sqrt{\tau} & 2\sqrt{\tau}\beta\\ 1 & 0\end{pmatrix}\begin{pmatrix}1 & x\\ 0 & 1\end{pmatrix})\psi'(x)dgdx$$

$$+\,c_1\int_{K_{F_{\mathbb{A}}}}\int_{F_{\mathbb{A}}^{\times}}\int_{E_{\mathbb{A}}} f'\left(k\begin{pmatrix}1 & x\\ 0 & a\sqrt{\tau}\end{pmatrix}\right)\psi'(x)|a|_{F_{\mathbb{A}}}^{-1}dkd^{\times}adx$$

where everything converges.

On the other hand, by Bruhat decomposition, the right hand side of (1) is convergent and equal to

$$\sum_{a\in F^{\times}}\int_{F_{\mathbb{A}}}\int_{F_{\mathbb{A}}} f\left(\begin{pmatrix}1 & x\\ 0 & 1\end{pmatrix}\begin{pmatrix}0 & a\\ 1 & 0\end{pmatrix}\begin{pmatrix}1 & y\\ 0 & 1\end{pmatrix}\right)\psi(x+y)dxdy + c_1\int_{F_{\mathbb{A}}} f\begin{pmatrix}1 & z\\ 0 & 1\end{pmatrix}\psi(z)dz.$$

Consequently, we need to verify that

$$\iint f'\left(g^{-1}\begin{pmatrix}\sqrt{\tau} & 2\sqrt{\tau}\beta\\ 1 & 0\end{pmatrix}\begin{pmatrix}1 & x\\ 0 & 1\end{pmatrix}\right)\psi'(x)dgdx = \iint f\left(\begin{pmatrix}1 & x\\ 0 & 1\end{pmatrix}\begin{pmatrix}0 & -b\\ 1 & 0\end{pmatrix}\begin{pmatrix}1 & y\\ 0 & 1\end{pmatrix}\right)\psi(x+y)dxdy$$

for $b = N_{E/F}(\beta)$;

$$\iint f\left(\begin{pmatrix}1 & x\\ 0 & 1\end{pmatrix}\begin{pmatrix}0 & -b\\ 1 & 0\end{pmatrix}\begin{pmatrix}1 & y\\ 0 & 1\end{pmatrix}\right)\psi(x+y)dxdy = 0$$

for $b \notin N_{E/F}(E^{\times})$; and

$$\iiint f'\left(k\begin{pmatrix}1 & x\\ 0 & a\sqrt{\tau}\end{pmatrix}\right)\psi'(x)|a|_{F_{\mathbb{A}}}^{-1}dkd^{\times}adx = \int_{F_{\mathbb{A}}} f\begin{pmatrix}1 & z\\ 0 & 1\end{pmatrix}\psi(z)dz.$$

We will treat these equalities locally.

At a place v of F which does not split in E, we can prove that

$$\psi'_w(\varphi)\int_{Z_{F_v}\backslash GL(2,F_v)}\int_{E_w} f'_w\left(g\begin{pmatrix}\sqrt{\tau} & 2\sqrt{\tau}\beta\\ 1 & 0\end{pmatrix}\begin{pmatrix}1 & x\\ 0 & 1\end{pmatrix}\right)\psi'_w(x)dgdx =$$

$$= \int_{F_v}\int_{F_v} f_v\left(\begin{pmatrix}1 & x\\ 0 & 1\end{pmatrix}\begin{pmatrix}0 & -b\\ 1 & 0\end{pmatrix}\begin{pmatrix}1 & y\\ 0 & 1\end{pmatrix}\right)\psi_v(x+y)dxdy$$

for $b = N_{E/F}(\beta) \in F_v^{\times}$;

$$\int_{F_v}\int_{F_v} f_v\left(\begin{pmatrix}1 & x\\ 0 & 1\end{pmatrix}\begin{pmatrix}0 & -b\\ 1 & 0\end{pmatrix}\begin{pmatrix}1 & y\\ 0 & 1\end{pmatrix}\right)\psi_v(x+y)dxdy = 0$$

for $b \in F_v^\times$ but $\notin N_{E_w/F_v}(E_w^\times)$; and

$$|\tau|_v^{-1/2}\int_{K_{F_v}}\int_{F_v^\times}\int_{E_w} f'_w\left(k\begin{pmatrix}1 & x\\ 0 & a\sqrt{\tau}\end{pmatrix}\begin{pmatrix}1 & x\\ 0 & 1\end{pmatrix}\right)\psi'_w(x)|a|_v^{-1}dkd^\times adx = \int_{F_v} f_v\begin{pmatrix}1 & x\\ 0 & 1\end{pmatrix}\psi_v(x)dx.$$

In fact, we may choose f_v by Satake transforms when f'_w is bi-invariant under K_w. At a place v of F which splits into w_1 and w_2 in E, we can also show the corresponding local equalities. In fact, we may choose

$$f_v(g) = f'^{\vee}_{w_2} * f'_{w_1}\left(\begin{pmatrix}1 & 0\\ 0 & -1\end{pmatrix}g\right)$$

where $f'^{\vee}_{w_2}(g) = f'_{w_2}(g^{-1})$.

By the Hasse Norm Principle and the facts that

$$\prod_v |\tau|_v = 1, \quad \prod_w \psi'_{w'}(\beta) = 1$$

where $\beta \in E^\times$, we have proved the proposition.

References

[1] P. Cartier, *Representations of p-Adic groups: A Survey*, Proceedings of Symposia in Pure Mathematics, Vol. 33 (1979), part 1, pp. 111-155.

[2] S. Gelbart and H. Jacquet, *Forms of GL(2) from Analytic Point of View*, Proceedings of Symposia in Pure Mathematics, Vol. 33 (1979), part 1, pp. 213-251.

[3] J. Jacquet and K.F. Lai, *A Relative Trace Formula*, Compositio Mathematica, 54 (1985), pp. 243-310.

[4] R.P. Langlands, *Base Change for GL(2)*, Annals of Mathematics Studies, 96 (1980).

[5] Yangbo Ye, Ph.D. Thesis, Columbia University, 1986.

Department of Applied Mathematics
Qing Hua University
Beijing

ABCDEFGHIJ — 898